Fire Safety in Residen Property

This book aims to take the reader through all aspects of fire safety and management in residential settings, from origin and ignition, risk assessment, protection and prevention, as well as comparing effective enforcement options from across all parts of the UK. It outlines the basis of law, standards and guidance relating to fire safety and building performance, and it critically evaluates the legal provisions and approaches to risk reduction with the focus on rented properties.

This book:

- Provides wider access to fire safety knowledge previously generally used by regulators and specialists.
- Examines fire risk assessments in domestic premises and the competency of assessors.
- Explains the approaches to fire safety enforcement and the impact of property licensing.
- Includes fire risk precautions for housing and general checklists to help landlords and tenants understand their responsibilities.
- Explores the effect of existing legislation with references to key Property Tribunal decisions relating to fire risk management and future legal developments.

This book will assist Environmental Health Officers and Environmental Health Practitioners – as well as graduating academics of the field – in their work to encourage the appropriate and effective use of legislation. Landlords, estate managers, student accommodation managers, surveyors and tenant groups may also find this book of interest.

Richard Lord is a Chartered Environmental Health Practitioner who has been a member of the Chartered Institute of Environmental Health since 1998. He is the owner and director of Housing Audit Services Ltd, an independent housing and health consultancy service providing support services to private sector landlords, managing agents and private sector providers of student accommodation. He specialises in fire safety in HMOs, assisting landlords and managers with overall compliance with the Housing Act 2004 and associated regulations, and fire risk assessments. Richard is part of the national team of ANUK/Unipol verifiers who audit members' compliance against the National Codes of Standards for Larger Residential Developments. He has particular expertise concerning fire safety in student accommodation. He provides HMO support services, including fire safety risk assessments, to landlords and portfolio managing agents of smaller HMOs mainly across the eastern portion of England.

Routledge Focus on Environmental Health
Series Editor: Stephen Battersby, MBE PhD, FCIEH, FRSPH

Pioneers in Public Health
Lessons from History
Edited by Jill Stewart

Tackling Health Inequalities
Reinventing the Role of Environmental Health
Surindar Dhesi

Statutory Nuisance and Residential Property
Environmental Health Problems in Housing
Stephen Battersby and John Pointing

Housing, Health and Well-Being
Stephen Battersby, Véronique Ezratty and David Ormandy

Selective Licensing
The Basis for a Collaborative Approach to Addressing Health Inequalities
Paul Oatt

Assessing Public Health Needs in a Lower Middle Income Country
Sarah Ruel-Bergeron, Jimi Patel, Riksum Kazi and Charlotte Burch

Fire Safety in Residential Property
A Practical Approach for Environmental Health
Richard Lord

Fire Safety in Residential Property

A Practical Approach for Environmental Health

Richard Lord

LONDON AND NEW YORK

First published 2021
by Routledge
2 Park Square, Milton Park, Abingdon, Oxon OX14 4RN

and by Routledge
605 Third Avenue, New York, NY 10158

Routledge is an imprint of the Taylor & Francis Group, an informa business

British Library Cataloguing-in-Publication Data
A catalogue record for this book is available from the British Library

Library of Congress Cataloging-in-Publication Data
A catalog record for this book has been requested

ISBN: 978-0-367-61784-4 (hbk)
ISBN: 978-0-367-61786-8 (pbk)
ISBN: 978-1-003-10656-2 (ebk)

Typeset in Bembo
by Apex CoVantage, LLC

Contents

Images

Primary legislation

Act for the Rebuilding of the City of London 1667
Building Act 1984
  Section 35
  Section 35A
  Section 36
Fire Precautions Act 1971
  s.6
  s.10
Fires Prevention (Metropolis) Act 1774 s.83
Fire (Scotland) Act 2005
Furniture and Furnishing Regulations 1988
Homes (Fitness for Human Habitation) Act 2018
Housing Act 1985 s.352
Housing Act 2004
  s.1(4)(d) and (5)
  s.10
  s.55(5)(c)
  s.64(3)
  s.65(1)
  s.67(4)
  s.90
  s.233
  s.234(s)
  sch. 1
  sch. 2
  sch. 14
The Housing (Wales) Act 2014
Public Health Act 1936 s.60
Worcester Ordinances 1467

Secondary legislation

Building Regulations 1991
Building (Approved Inspectors etc.) Regulations 2000
Building (Scotland) Regulations 2004
Building Regulations 2010 s.38
Building Regulations (Northern Ireland) 2012
Electrical Safety Standards in the Private Rented Sector (England) Regulations 2020
Fire and Rescue Services (Northern Ireland) Order 2006
Fire Precautions (Hotels and Boarding Houses) Order 1972, Statutory Instrument 1972 No. 238
Fire Precautions (Hotels and Boarding Houses) (Scotland) Order 1972, Statutory Instrument 1972 No. 382 s.26
Fire Safety (Scotland) Regulations 2006 SI 456

Fire Safety Regulations (Northern Ireland) 2010 SI 325
Furniture and Furnishing Regulations 1988
General Product Safety Regulations 2005
Housing Health and Safety Rating System (England) Regulations 2005 SI 3208
Housing Health and Safety Rating System (Wales) Regulations 2006 SI 1702 (W164)
Houses in Multiple Occupation (Specified Educational Establishments) (England) Regulations
Houses in Multiple Occupation (Specified Educational Establishments) (Wales) Regulations 2006 No. 1707 (W169)
Licensing and Management of Houses in Multiple Occupation and Other Houses (Miscellaneous Provisions) (England) Regulations 2006 No. 373
Licensing and Management of Houses in Multiple Occupation (Additional Provisions) (Wales) Regulations 2007 (SI 2007/3229) (W281)

Management of Houses in Multiple Occupation (England) Regulations 2006 No. 372
reg. 4(1)
reg. 4(2)
reg. 10
Management of Houses in Multiple Occupation (Wales) Regulations 2006 No. 1713 (W175)
Regulatory Reform (Fire Safety Order) 2005
Article 4
Article 8
Article 29
Article 30
Article 31
Article 31(6)
Article 43
Selective Licensing of Houses (Specified Exemptions) (England) Order 2006 SI 370
Smoke and Carbon Monoxide Alarm (England) Regulations 2015

British Crown dependencies and Australia

Isle of Man

Fire Precautions (Flats) Regulations 1996
The Fire Precautions (Houses in Multiple Occupation and Flats) Regulations 2016 – Statutory Document No. 2016/0218

Jersey

Fire Precautions (Jersey) Law 1977
Fire Precautions (Designated Premises) (Jersey) Regulations 2012

Australia

New South Wales Government – Boarding Houses Act 2012 No. 74
Queensland Government – Building Fire Safety Regulations 2008

Glossary of acronyms

ACM	Aluminium Composite Material (the cladding used on Grenfell Tower and Lacrosse Tower)
BSA	Building Societies Association
BSB	Building Safety Bill
CSG	Competence Steering Group
EHO	Environmental Health Officer
EHP	Environmental Health Practitioner
EICR	Electrical Installation Condition Report
EWFR	External Wall Fire Review
FRA	Fire Risk Assessment (must be 'suitable and sufficient')
FSB	Fire Safety Bill
FSO	Regulatory Reform (Fire Safety) Order 2005
F-tT	First-tier Tribunal (Property Chamber) (handles applications, appeals and references relating to disputes over property and land)
GPSR	General Product Safety Regulations 2005
HEI	Higher Education Institution (a term from the Further and Higher Education Act 1992, including universities and colleges)
HHSRS	Housing Health and Safety Rating System
HMO	House in Multiple Occupation
HPL	High Pressure Laminate (the cladding used on the Cube student accommodation, Bolton)
HRRB	High-rise Residential Buildings
HSE	Health and Safety Executive
IFSSC	International Fire Safety Standards Coalition
LACORS	Local Authorities Coordinators of Regulatory Services (organisation no longer exists; published *Housing – Fire Safety: Guidance on Fire Safety Provisions for Certain Types of Existing Housing*)

LHA	Local Housing Authority
MHCLG	Ministry of Housing, Communities and Local Government
NFSCC	National Fire Safety Chiefs Council
NHBC	National House Building Council
PBSA	Purpose-built Student Accommodation
PEEP	Personal Emergency Evacuation Plan
RICS	Royal Institution of Chartered Surveyors
RPT	Residential Property Tribunal (operates in Wales)
UT	Upper Tribunal (Lands Chamber) (responsible for handling appeals against decisions made by First-tier Tribunal (Property Chamber) and Residential Property Tribunal in Wales)

Acknowledgements

The idea of writing a monologue started in Autumn 2019 from a brief conversation with John Pointing on a train journey returning from Leeds, where we had been providing training together. I had never considered the idea of writing a book, so thank you, John, for your confidence and encouragement. In March 2020, as a private consultant, the COVID-19 lockdown left me with the prospect of no work and nothing to do, so it seemed an ideal time to try and produce this book. Further thanks go to Dr Stephen Battersby, who suggested the topic of fire safety, as he knows it is a passion of mine in which I have invested a considerable amount of time developing my knowledge.

Further thanks go to Martin Blakey, Richard Pixner, Martin Rushall, Richard Tacagni and Simon West for their encouragement and assistance.

Finally I must thank my wife, Hazel, and my children, Esme and Mischa, for letting me get on with it at times when there were other, more fun things to do.

1 Introduction

1.01

In recent years, a number of major fires in high-rise tower blocks have produced renewed public and professional concern about fire safety, not least because of the high number of fatalities resulting from landmark incidents such as those at Lakanal House (2009) and Grenfell Tower (2017) in London. Added to these, other non-fatal fires – including the Cube in Bolton (student accommodation), Samuel Garside House and Barking (both in 2019) and Lacrosse Tower in central Melbourne, Australia (2014) – have raised awareness of the devastating effects of residential fires on the physical and mental health of survivors and, more generally, on the fabric of local communities. According to the UK Home Office, at incidents attended by the Fire and Rescue Service, there are more than 7,000 non-fatal casualties in England each year.[1] In Scotland, there were over 1,100 in 2017–18.[2] These high numbers are partly a function of inadequate regard for fire safety in many new-build family houses.[3]

1.02

Although, overall, the UK continues to suffer from a systemic undersupply of new housing, the size of the private rented sector has increased dramatically over the past ten years. Rising demand has been driven in particular by:

- an acknowledged deficit in affordable housing;
- widening participation in the marketised higher education sector, resulting in rapid expansion in full-time student intakes and purpose-built accommodation to house these students;
- a growing numbers of young adults and migrants looking to establish themselves in shared housing or bedsit conversions.

The current UK Conservative Government has committed to bring into commission one million new homes by the end of 2020 and to 'deliver half a million more by the end of 2022.'[4] The scale of this ambition requires not only more rapid building but also, in many cases, the use of cheaper materials, for example, lightweight timber frames and other modern methods of construction (MMC).[5] It is generally accepted that effective fire safety prevention and management for all types of residential settings require a joined-up multi-agency approach, in which a range of professions collaborate and co-operate to make sure a building is correctly built, managed and maintained. Planners, builders, building control officers, fire risk assessors and building managers must work towards the goal of delivering homes that are safe to live in. However, occasionally, badly built or refurbished residential buildings that are certified as meeting building regulations subsequently burn down.

1.03

Occupiers have a responsibility to minimise fire hazards in their homes. To support householders across the UK, the National Fire Safety Chiefs Council (NFSCC) coordinates and supports a number of national safety week initiatives every year. These include Home Safety Week, the main purpose of which is to encourage occupiers to make sure they have sufficient fire alarms in their homes and to test them. According to the NFSCC 'in 20 per cent of accidental dwelling fires in the UK, smoke alarms fail to activate.'[6]

1.04

In England and Wales there is a dual enforcement approach to fire safety in residential buildings, which can be confusing to landlords, tenants and building managers. Local Housing Authority (LHA) officers – typically Environmental Health Officers (EHOs) – assess the fire hazards in residential premises using the Housing Health and Safety Rating System (HHSRS) under the Housing Act 2004 and associated regulations. The purpose of these assessments is to minimise the risk across all tenures of housing through advice, improvements and property licensing conditions. Where dwellings fail to meet fire safety standards of refurbishment or change of use – for example, from a family house to a House in Multiple Occupation (HMO) – it is often EHOs who assess and take steps to minimise fire hazards in partnership with fire and rescue authorities. However, EHOs increasingly require additional knowledge to assess the actual risk of fire across all parts of residential developments. As well as the individual dwellings, assessment includes the internal common parts and external parts, particularly of high-rise residential buildings (HRRBs). In the UK, housing as a government function is devolved across the four constituent nation states and is characterised by different regulatory approaches. The self-governing Crown dependencies (Jersey, Guernsey and the Isle of Man) also have independent approaches to fire safety regulation.

1.05

The risk of a domestic fire starting and the damage it can cause vary according to the specific circumstances. Depending on the seriousness of a fire's impact, questions may be asked in inquests, inquiries, court cases or risk assessment reviews to try to understand what went wrong and why, and how to prevent a recurrence. Like much other health and safety law around the world, UK fire safety legislation has developed and evolved over centuries, often in direct response to specific tragedies and with a view to minimise the likelihood of a recurrence. Chapter 2 explores how the UK's body of fire safety law has been shaped from before the Great Fire of London to the present day, considers legislation in other countries that is relevant to fire prevention in the home and provides a focus on historic disasters from which major fire safety legislation was developed.

1.06

Today a high level of specialised knowledge and professional expertise is often necessary in assessing risks in specific buildings, not least large and complex residential dwellings. EHOs and non-specialists in fire safety inevitably have variable levels of understanding of the fire hazard and risks in domestic settings, and often they are unable to get advice and guidance from fire and rescue authorities as a consequence of funding cuts and under-resourcing. As a basis for developing further knowledge, Chapter 3 sets out the principles of fire, specifically in a domestic setting. These include ignition sources, fire formation and compartmentation to prevent spread and what fire precautions may be required, based on the risk.

1.07

Landlords are required to make sure that a 'suitable and sufficient' fire safety risk assessment is made of the common parts of blocks of flats and HMOs to protect life safety, especially in sleeping accommodation.[7] There is no legal requirement for a risk assessment to be undertaken by a third-party professional. However, many landlords and building managers engage consultants to act on their behalf, although they (the landlords or managers) remain responsible for the suitability of assessment. Landlords need to be able to determine what level of fire risk assessment is necessary for a particular building. Also, they should have sufficient knowledge either to undertake the assessment themselves or to adequately instruct an independent assessor who is sufficiently knowledgeable to undertake it on their behalf. Landlords need to be aware of the consequences of not engaging a competent assessor. Many risk assessors are not familiar with the HHSRS and the relationship with the Regulatory Reform (Fire Safety Order) 2005 (FSO). Landlords of smaller properties may only require a basic understanding of ignition, fuel, fire protection measures and what routine management is necessary to allow them to undertake a risk assessment themselves. Chapter 4 explores fire risk assessments and the competency of assessors, fire service enforcement and, briefly, the impact of the building regulations on the fire risk. Chapter 5 considers levels of fire safety management, including evacuation strategies, testing and maintenance of fire alarms, fire extinguishers and fire doors and the management of electrical safety.

1.08

In recent years, there has been a rapid rise in the number of property licensing schemes operating in the UK, particularly in larger cities and university towns. In this context, there has never been a greater need for all professions that have responsibility to manage, risk assess and regulate housing to fully understand how to use the differing parts of the Housing Act 2004 correctly and how the relationship with the FSO should work. LHA property licensing schemes can impose conditions on licence holders, requiring them to provide fire precautionary facilities and effective management of privately rented houses and HMOs. The importance of correct use of the different parts of the Housing Act 2004 property licensing schemes will be explored in Chapter 6.

1.09

As a result of local authority funding cuts, EHOs rarely have the capacity to offer free advice and to prepare extensive schedules of work to help landlords and managers meet their fire safety obligations. In better-resourced times, EHOs would, for example, have been able to advise on larger HMOs or converted blocks of flats with multiple cooking facilities and poor separation. Landlords are increasingly expected to meet their responsibilities themselves or to engage a third-party consultant to help them. These issues are dealt with in Chapter 7, along with references to the *LACORS Housing – Fire Safety: Guidance on Fire Safety Provisions for Certain Types of Existing Housing*, which is considered by LHA officers and First-tier Tribunals as the gold standard reference documentation for fire safety in England. The management and risk assessment of high-rise residential buildings will be considered in Chapter 8. This will include the hazard presented by external wall cladding and licensing of larger blocks of student accommodation. The final chapter will consider the future legal and fire safety policy developments and Government responses arising out of the Grenfell Tower Inquiry, still in session at the time of writing. These responses will include proposed new legislation and formal recognition that fire risk assessors should be competent.

1.10

This work differs from other fire safety and risk assessment publications in its exclusive focus on the hazard of fire in residential settings. It aims to bring together current available information into one concise publication to help LHA officers, landlords, property managers and student accommodation managers to understand the relationship between the Housing Act 2004 and the FSO and to use the correct legislation appropriately and effectively.

Notes

1 The Home Office collects detailed information on incidents attended by fire and rescue services, www.gov.uk/government/collections/fire-statistics
2 See Scottish Fire and Rescue Service, www.firescotland.gov.uk/about-us/fire-and-rescue-statistics.aspx
3 See *New-build Homes Not Fire Safe*, BBC News, 30 April 2019, www.bbc.co.uk/news/business-48113301
4 *Housebuilding Targets* – Debate Pack Number CDP 2019–0147, House of Commons Library, 10 June 2019
5 See *Modern Methods of Construction: Who's Doing What?* NHBC Foundation, November 2018, www.nhbcfoundation.org/wp-content/uploads/2018/11/NF82.pdf
6 See National Chiefs Safety Council, www.nationalfirechiefs.org.uk/Home-Safety-Week
7 See the Regulatory Reform (Fire Safety) Order 2005 Article 9(1): Risk assessment

2 The development of fire safety legislation

2.01

Historically, devastating fires have prompted changes to fire safety legislation. This is as true of the UK Parliament as it is of legislatures around the world. Like other health- and safety-based law-making, fire safety legislation has often been a response to a tragic event rather than the output of an informed, risk-based analysis of a hazard. This has produced a pattern of legislation that is reactive, piecemeal and *ad hoc*. As such, it does not always keep pace with changes to lifestyle and construction standards that can alter fire risk.

2.02

At the time of writing, the UK Government has launched a fire safety consultation across England, following the Grenfell Tower fire, as *'[I]t shook confidence in the building safety system to the core. . . . As a Government we are determined to learn lessons.'*[1] This statement is an echo of a parliamentary debate in 1971 that recognised the disjointed approach to fire safety law in the UK:

> *The laws concerning fire precautions have developed piecemeal to meet different needs which have arisen from time to time in the past. They are applied by different authorities for different purposes.*[2]

2.03

In the same debate it was recognised that the development of fire safety law has generally been reactive and not always straightforward for property owners, managers and tenants to understand, with overlaps and confusion possibly contributing to further tragedies:

It is a fact that all too often in the past before action has been taken it has needed some major catastrophe to focus attention on some weakness or other in our provisions for the safety of life in the event of fire. It is fair to say that the law relating to means of escape and other fire precautions has evolved over the years to meet different needs at different times.

The third deficiency in the law as it stands is the accumulation of separate laws, both general and local, and the different ways in which they are enforced. There is no doubt that this leads to confusion. Under the existing provisions the owners and occupiers of property do not know, and cannot be expected to know, what is expected of them.[3]

2.04

It is still the case that two major pieces of overlapping fire-related legislation in England are enforced by different authorities (fire and rescue authorities and LHAs), making it far from easy for landlords and managers to fully understand their responsibilities. This can lead to accidental non-compliance by landlords as a consequence of ignorance rather than wilful negligence. The current fire safety legislation across England and Wales is largely risk-based rather than prescriptive. It is supported by non-statutory guidance documents to assist those responsible for fire safety.

2.05

In Britain, the history of fire safety regulation and control in buildings can be traced back several centuries. Historically, the system of local government required local bylaws to be passed to address local problems. The **Worcester Ordinances of 1467** introduced compulsory tiling on roofs to reduce the number of houses with thatched roofs in order to minimise the spread of fire.[4]

2.06

Set out next a brief history of the development of fire safety laws and regulations with reference to some of the more significant fires that shaped that development. The Grenfell Tower Inquiry has led to a review of the Building Regulations 2010, which will result in legislative changes. These are discussed further in Chapter 9. Fire safety legislation continues to develop but is not a cure-all that will prevent and protect occupiers against fires.

2.07

Probably the earliest significant fire that many people will be familiar with is the 1666 Great Fire of London, which destroyed thousands of wooden thatched buildings in the City of London. In all, 13,200 houses, 87 churches and 52 livery company halls were destroyed, along with a number of courts, jails and civil administration buildings. The **Act for the Rebuilding of the City of London** was passed in February 1667 as a piece of local legislation, common in that period. Foremost among its provisions was the requirement that all new buildings were to be constructed of brick or stone. It imposed a maximum number of storeys per house and limited the number of people per property to reduce the chances of overcrowding.[5] This was an early example of local building regulation used to protect building safety.

2.08

The **Fires Prevention (Metropolis) Act 1774** required houses primarily in the cities of London and Westminster to be divided into separate classes of use, each with its own specification for wall thickness. It also required parishes to provide fire ladders as a direct way to protect life safety rather than property safety.[6] In England, Wales and Scotland, even today a tenant of a property destroyed by fire can invoke Section 83 of the 1774 Act in order to reinstate the damaged building. The Act, however, has been removed from the statute books in New Zealand, where it ceased to have effect on 1 January 2008.

2.09

For blocks of flats or tenements that were more than two storeys high and where the upper storey was more than 20 feet above the surface of the street or ground on any side of the building, the **Public Health Act 1936, s.60** required the provision of a means of escape in the event of fire. It was recognised that most accidental fires started in the home and that a fire in a block of flats had the potential to affect many families. This legislation gained increased importance after World War II, when mass house building and slum clearance concentrated on providing affordable homes that were often in high-rise blocks of flats. This section was later incorporated into the Building Regulations.

2.10

The 11 deaths caused by a fire at the Rose and Crown Hotel, Saffron Walden, Essex on 26 December 1969 led to Parliament passing the first dedicated fire safety legislation, **the Fire Precautions Act 1971**.[7] In

England and Wales, statutory instruments gave the Secretary of State powers to designate specific types of premises that required fire certificates issued and enforced by the fire and rescue service.[8] Similar provisions were made for Scotland.[9] Only boarding houses (occupied by more than six people sleeping above the first floor or below the ground floor) and hotels were so designated. Successive Secretaries of State failed to add blocks of flats to the list of designations with the result that older blocks such as Grenfell Tower never required a fire certificate. The certification detail was specified in section 6 of the Act and included the use of the building, the means of escape, the type, number and location of fire-fighting equipment and the type of means of early warning. The certificates did not impose any responsibilities on the owners. Importantly, there was no requirement for owners to consider the assessment of the fire risk, for example, sources of ignition, fuel and any other factors that could lead to fire.

2.11

The fire authority had powers to serve a prohibition order under section 10 of the Act, if they were satisfied that there was an excessive risk to people in the event of fire. In order to secure authority to prohibit or restrict use of certain premises, the relevant statutory agencies in England made a complaint or in Scotland made a summary application to the courts. Interestingly, the courts were given the power to close a business based on the recommendation of the Fire Authority.[10]

2.12

The **Furniture and Furnishing Regulations 1988** are a set of UK-wide regulations that set levels of fire resistance for coverings and fillings in domestic upholstered furniture. They were introduced as a direct result of a serious fire on the second storey of Woolworth's, Manchester, which claimed ten lives in May 1979. Although fire-certificated, the building did not have a sprinkler system or smoke detectors, as these were discretionary at the time.

2.13

Investigation into the cause of the fire was inconclusive. Conflicting reports suggested that either a damaged electrical cable or a discarded cigarette was the cause of ignition. What *is* known is that polyurethane sofas were stacked on their ends and, once ablaze, burned at 700^0C, releasing a deadly cyanide gas. The high volume of toxic smoke produced quickly obscured the fire exits.[11] The Joint Fire Prevention Committee's report prompted a number of Home Office actions.[12]

2.14

In March 1980, following the coroner's inquest, the dangers posed by foam-filled furniture were the subject of a parliamentary debate that led to legislation requiring that every item of foam-filled furniture be protected by a fire-retardant coating and carry a warning label.[13] Research undertaken by the University of Surrey has indicated that the regulations helped save at least 710 lives by 1997.[14] All furniture provided in rented sleeping accommodation must meet the regulations and carry a label. Landlords who provide furnished accommodation are considered a distributor of furniture, because they are professional within the supply chain.

2.15

The **Housing Act 1985** empowered LHAs to award discretionary renovation grants to make general improvements to unfit houses and HMOs in disrepair but not unfit. Such general improvements included fire safety improvements. Section 352 of the Act also gave LHAs in England and Wales the power to improve means of escape and other fire precautions to make HMOs safe for the number of occupiers. Accompanying fire safety guidance was produced in 1992.[15] There was, however, no requirement for LHAs to follow the guidance.

2.16

The definition of an HMO was challenged in *Barnes v. Sheffield City Council* in 1995.[16] The court of appeal held that five students living together formed a single household, as opposed to residents living in an HMO. Sheffield City Council required the owners of a house occupied by five female students to install fire escapes and other fire precautions. The owners refused, claiming the house was not an HMO but was occupied as a single household. This judgement left LHAs powerless to require fire safety improvements in many HMOs, particularly shared student properties at that time. Landlords were able to avoid having to comply with fire safety requirements for HMOs if all the occupiers signed a joint and several tenancy agreement, regardless of how the occupiers used the property in reality.

2.17

By the early 2000s, UK-wide fire safety legislation was covered by approximately 70 different pieces of legislation.[17] On 1 October 2006, the Fire Precautions Act 1971 was repealed in England and Wales and replaced by the **Regulatory Reform (Fire Safety) Order 2005** in order to consolidate and simplify fire safety regulations. Similar regulations were introduced in Scotland[18] and Northern Ireland.[19] The Order was designed to improve overall fire safety in non-domestic settings, including hotels, hostels, mobile homes on permanent sites and the common parts of HMOs. However, the common parts of shared houses are considered as domestic premises and are therefore outside the scope. The most significant provision to protect sleeping occupiers is Article 9, which requires a responsible person to undertake a fire risk assessment, including in a building where a licence is in force. The Order was introduced before HMO licensing began in England. However, an unintended consequence is that HMO licence holders for houses that may be considered as lower-risk, shared-house arrangements are now legally required to undertake a suitable and sufficient fire risk assessment. This will bring many more shared houses into scope since the definition of HMOs that require a licence was extended in 2018.

2.18

The **Fire Safety Order** is enforced by the fire and rescue authorities. Where LHAs impose fire safety conditions for the common parts of an HMO licence in a shared HMO, they must enforce those conditions. However, any licence conditions applied to the common parts of an HMO where the Fire Safety Order applies shall have no effect under Article 43. This overlap may cause confusion for landlords and operators of larger blocks of student accommodation. It is therefore important that clear working relationships are agreed between both authorities and are regularly reviewed.

2.19

Introduced in England and Wales in 2006, the **Housing Act 2004** repealed large parts of the Housing Act 1985, including the rigid concept of the fitness standard. The new Act provided for an evidence-based system for housing inspections and enforcement, based on the risk of ill health to occupiers and guests, caused by hazards. This is known as the Housing Health and Safety Rating System (HHSRS), although, interestingly, the term is not used directly in the Act. The HHSRS does not apply in Scotland.

2.20

Fire is a prescribed hazard under s.10 of the Act for the purposes of HHSRS, which gave LHAs the responsibility to improve fire safety across housing of all tenures (initially excluding tenants of Council-owned properties) in conjunction with the fire and rescue services. The application of the HHSRS in dwellings occupied by LHA tenants in England was extended by the **Homes (Fitness for Human Habitation) Act 2018**. The practical use of HHSRS is discussed in Chapters 6 and 8.

2.21

The Act tightened regulation of HMOs by defining a household. One effect was to bring all shared houses into scope, thus increasing LHA regulation. Mandatory HMO licensing was also introduced in the UK.

2.22

The **Smoke and Carbon Monoxide Alarm (England) Regulations 2015** placed a duty on all private sector landlords to provide at least one smoke alarm on each storey at the beginning of each tenancy. A Government review of conditions in the rented sector in 2014 recognised that working smoke alarms saved lives and that there was, anomalously, no duty on private sector landlords to install a working smoke alarm except in licensed HMOs.[20] Local authorities were able to use HHSRS to require the installation of smoke alarms in any private dwelling, but many rented houses did not have smoke alarms fitted despite the risk. The regulations do not place any ongoing duty on the landlord to test the alarms, a responsibility that therefore defaults to the tenant. In November 2020, the Government published a consultative document, which proposes extending the requirements to social landlords and amending the statutory guidance to support Part J of the Building Regulations.[21]

Building Regulations

2.23

The system of building control across the UK requiring residential dwellings to be built to a satisfactory standard has evolved over many centuries.

The Building Act 1984 was the first national Act with central control that allowed the Government to issue Building Regulations across England and Wales.[22] Similar regulations covered Scotland[23] and Northern Ireland.[24] Approved documents are issued under the Building Regulations 2010. These provide general guidance and practical solutions on ways to meet the regulations. However, they do not prescribe a single means of achieving compliance. Dwellings and other buildings must meet the building control standard, but it is for the builder or designer to demonstrate to the inspector who approves the work how compliance will be achieved. Approved Document B in England is split into two volumes: volume 1 for dwellings and volume 2 for buildings other than dwellings (for example, blocks of flats, sheltered housing and student accommodation).

2.24

In all four of the UK's constituent nations, there is a requirement that:

- at least one interlinked working smoke alarm is installed on each level of a house;
- a heat alarm is installed in the kitchen in certain circumstances. Whether or not this requirement applies depends on the kitchen's separation from the hallway. The regulations cannot be enforced retrospectively, which means that some older owner-occupied houses may not have smoke alarms.

2.25

Section 38 of the Building Regulations 2010 requires the person carrying out the building works to provide fire safety information to the responsible person, when the building work is completed, in order to allow a fire safety risk assessment to be undertaken.[25]

2.26

The **General Product Safety Regulations 2005** (GPSR) treat landlords and managing agents who provide products (either new or second-hand) as 'distributors' or 'professionals within the supply chain.' Products that are not covered by other regulations (such as carpets and curtains) fall within the scope of the GPSR, along with inherited products. These must be safe for consumers when used in a normal or reasonably foreseeable way. Every year around 50 per cent of accidental fires in England are caused by faulty appliances and leads or electrics or by the misuse of appliances and leads.[26] Many of these are preventable; simple checks and proper instructions given to tenants can reduce the risk of ignition.

2.27

The **Electrical Safety Standards in the Private Rented Sector (England) Regulations 2020** came into force on 20 June 2020.[27] In rented homes, fixed electrical installations (including wiring, sockets and the fire resistance of consumer unit enclosures) must be inspected and tested every five years to bring the sector into line with electrical inspection tests for HMOs. These regulations will help to further reduce fires and electrical injuries caused by electrical faults.

Australia

2.28

Australian Acts that are passed by either State Parliament or the Commonwealth Parliament must receive Royal Assent from the Queen, as head of state, through her representative. Australian common law mirrors the English common law system and has developed by the consideration of decisions made by Australian superior courts and relevant English courts over many years.[28]

Australian laws are both federal laws covering the entire population and state laws covering people in a particular state or territory. State laws only apply to that individual state's jurisdictional area, meaning that states can approach fire safety as they see fit. As an example, residential smoke alarms are mandatory in Victoria and South Australia, but they used not to be in New South Wales (NSW). However, NSW decided to follow their neighbours' lead, when, over a two-week period in 2005, a spate of fires claimed 13 lives in the state.[29] Now, everyone who lives in NSW must have smoke alarms fitted in every storey of their house or dwelling (apartments, hostels and hotels) and in rooms where people sleep.

Across Australia, there are different approaches to fire safety enforcement in private rented accommodation. Local council officers conduct random fire inspections of boarding houses to enforce fire laws in NSW[30] under the Boarding Houses Act 2012.[31] These fire inspections may be unannounced, depending on the individual circumstances of the case. By contrast, in Queensland the fire and rescue services enforce the Building Fire Safety Regulation 2008.[32]

Isle of Man

2.29

The Fire Precautions (Houses in Multiple Occupation and Flats) Regulations 2016 were introduced on 1 January 2017.[32] Before this date there had been no fire safety standards specified for HMOs on the island.[33] There is no requirement to make any fire safety improvement to flats that were inspected and signed off as complying with the Fire Precautions (Flats) Regulations 1996 before January 2017. However, material changes to flats must meet the current regulations, which were introduced to meet modern standards.[34] Inspections of flats and HMOs are carried out by the fire service, and officers are available to provide advice.

The States of Jersey

2.30

States of Jersey fire safety law was originally modelled on UK legislation, but it remained unchanged for a long period, dating from 1977. An internal review in 2010 recognised the need for changes to modernise fire safety in Jersey by learning from the changes that had been made in the UK and Europe.[35] A series of amendments were made to the Fire Precautions (Jersey) Law 1977. Introduced on 1 January 2013, these included the move to make fire certification a requirement for the majority of private rented accommodation across the States of Jersey, including hostels, lodging houses and HMOs (also redefined). Single private houses remain exempted from the certification process. Applications for fire certificates are made to the States of Jersey Fire and Rescue Service, which is also responsible for enforcement.[36] Fire certificates must be renewed every 36 months. The list of specific premises that require fire certificates is set out in the Fire Precautions (Designated Premises) (Jersey) Regulations 2012.

2.31

When renewing fire certificates, applicants must advise the fire service of any material alterations made to the certified premises. These include alterations to the escape routes and other relevant fire precautions involving premises that would be inadequate when used in normal conditions. The Chief Fire Officer must also be notified in advance of any plans to make material alterations to the certified premises.[37]

2.32

Owners of HMOs are required to provide floor plans to the fire service, which determines the necessary works required by way of a notice. Provided the works completed are of a satisfactory standard, the fire service issues a fire certificate once it has completed its inspection.[38]

Notes

1 See Home Office, *Fire Safety, Government Consultation*, July 2020, Foreword, https://assets.publishing.service.gov.uk/government/uploads/system/uploads/attachment_data/file/919566/20200717_FINAL_Fire_Safety_Consultation_Document.pdf

2 The Minister of State, Home Office (Lord Windlesham), Fire Precautions Bill debates, Hansard, 18 March 1971, Volume 316, https://hansard.parliament.uk/Lords/1971-03-18/debates/dafb326e-25f9-4b94-b847-311a3fd8d1ab/FirePrecautionsBill?highlight=fire%20precautions%20bill#contribution-5cd806fc-6b53-444b-bb97-17fee339c795

3 See Hansard debate, https://hansard.parliament.uk/Commons/1970-11-20/debates/4dc904ce-29b4-4898-b5b3-7f1047b6cb56/FirePrecautionsBill?highlight=fire%20precautions%20bill#contribution-20bbd4a3-b4ec-4139-8962-209a18446652

4 British History Online, *The City of Worcester: Introduction and Borough*, www.british-history.ac.uk/vch/worcs/vol4/pp376-390#highlight-first

5 UK Parliament, *An Act for Rebuilding the City of London*, 2020, www.parliament.uk/about/living-heritage/transformingsociety/towncountry/towns/collections/collections-great-fire-1666/1666-act-to-rebuild-the-city-of-london/

6 See Bishop and Sewell LLP, *Property Law: The Fires Prevention (Metropolis) Act of 1774; A 246-year-old Law Still on the Books and Able to be Utilised in our Modern Times*, 2020, www.bishopandsewell.co.uk/2020/02/17/property-law-the-fires-prevention-metropolis-act-of-1774-a-246-year-old-law-still-on-the-books-and-able-to-be-utilised-in-our-modern-times/

7 Saffron Walden Reporter, *Fatal Hotel Fire in Saffron Walden Remembered 50 Years On*, 8 February 2019, www.saffronwaldenreporter.co.uk/news/firefighters-remember-rose-and-crown-hotel-fire-in-saffron-walden-50-years-on-1-5883958

8 Fire Precautions (Hotels and Boarding Houses) Order 1972, Statutory Instrument 1972 No. 238

9 Fire Precautions (Hotels and Boarding Houses) (Scotland) Order 1972, Statutory Instrument 1972 No. 382 (s.26)

10 Fire Precautions Act 1971, Chapter 40, s.10, www.legislation.gov.uk/ukpga/1971/40/pdfs/ukpga_19710040_en.pdf

11 See Manchester Evening News, www.manchestereveningnews.co.uk/news/greater-manchester-news/toxic-smoke-filled-woolworths-windows-15156337

12 www.woolworthsmuseum.co.uk/1970s-fire.htm

13 See Hansard, *Woolworth's Store Manchester Fire*, 18 March 1980, https://hansard.parliament.uk/Commons/1980-03-18/debates/873c1062-cc90-44ea-8799-9b35d12d5775/WoolworthSStorePiccadillyManchester(Fire)?highlight=woolworths%20fire%20manchester#contribution-74bae8fc-a8fc-47d4-a944-688291018dbf

14 The UK Government Department for Trade and Industry commissioned the University of Surrey to research the effectiveness of the Furniture and Furnishings (Fire) (Safety) Regulations 1988. The resulting report – *Effectiveness of the Furniture and Furnishings (Fire) (Safety) Regulations 1988* – was published in June 2000, www.satra.com/spotlight/article.php?id=382
15 Department of Environment Circular 12/92, *Houses in Multiple Occupation – Guidance to Local Housing Authorities on Standards of Fitness under s.352 of the Housing Act 1985* (This guidance has been withdrawn)
16 See Hansard debate on University Students (Accommodation), 10 February 1999, https://api.parliament.uk/historic-hansard/commons/1999/feb/10/university-students-accommodation
17 See Fire Safety Advice Centre – Regulatory Reform Order 2005, www.firesafe.org.uk/regulatory-reform-fire-safety-order-2005/
18 Fire Safety (Scotland) Regulations 2006, SI 456
19 Fire Safety Regulations (Northern Ireland) 2010, SI 325
20 Department for Communities and Local Government, Review of Property Conditions in the Private Rented Sector, February 2014
21 Ministry of Housing Communities and Local Government, *Open Consultation – Domestic Smoke and Carbon Monoxide Alarms: Proposals to Extend Regulations Updated 24 November 2020*, www.gov.uk/government/consultations/domestic-smoke-and-carbon-monoxide-alarms/domestic-smoke-and-carbon-monoxide-alarms-proposals-to-extend-regulations
22 SI 2010. No 2214
23 The Building (Scotland) Regulations 2004
24 The Building Regulations (Northern Ireland) 2012
25 Regulations 38(3)(a): 'Fire safety information' means information relating to the design and construction of the building or extension, and the services, fittings and equipment provided in, or in connection with, the building or extension which will assist the responsible person to operate and maintain the building or extension with reasonable safety.
26 Fire Statistics Table 0601: Primary fires in dwellings and other buildings, by cause of fire, England
27 SI 312
28 See Caxton Legal Centre Ins Australian Common Law, https://queenslandlawhandbook.org.au/the-queensland-law-handbook/the-australian-legal-system/where-law-comes-from/common-law/
29 www.cityofsydney.nsw.gov.au/business/doing-business-with-us/regulations/health-and-safety/fire-safety
30 City of Sydney, *Fire Safety Guidelines for Boarding Houses*
31 New South Wales Government, Boarding Houses Act 2012 No. 74
32 Fire Precautions (Houses in Multiple Occupation and Flats) Regulations 2016
33 Carl Kinvig, Senior Fire Safety Officer: 'Previously there were no fire safety standards specified for houses in multiple occupation, which present a similar fire risk as flats. We have worked closely with DEFA [Department of Environment, Food and Agriculture] to modernise the legislation as part of our commitment to protecting the community.' www.gov.im/news/2017/jan/17/new-fire-regulations-in-operation/
34 Guidance to the Fire Precautions (Houses in Multiple Occupation and Flats) Regulations 2016

35 Fire Precautions Law and Designated Premises Regulations, www.gov.je/government/consultations/pages/fireprecautionslawdesignatedpremisesregulations.aspx
36 Fire Precautions (Jersey) Law 1977, Article 2A: Transitional arrangements
37 Government of Jersey, www.gov.je/StayingSafe/FireSafety/Workplace/Legislation/Pages/FireCertificates.aspx#:~:text=The%20Fire%20Precautions%20(Jersey)%20Law%201977%20requires%20that%20the%20occupier,inspection%20as%20and%20when%20required
38 Jersey Evening Post, *Hundreds in Island Flats with no Fire Certificate*, 15 June 2017, https://jerseyeveningpost.com/news/2017/06/15/hundreds-in-island-flats-with-no-fire-certificate/

Further reading

A.J. Ley, *Building Control UK – An Historical Review*, www.irbnet.de/daten/iconda/CIB1598.pdf

London Fire Brigade, *Fire of London*, www.london-fire.gov.uk/museum/history-and-stories/the-great-fire-of-london/

Parliament of Australia, *Use of Smoke Alarms to Prevent Smoke and Fire Related Deaths*, April 2016: Chapter 2, *Smoke and Fire Related Incidents, Deaths, Injuries and Property Damage*, www.aph.gov.au/Parliamentary_Business/Committees/Senate/Legal_and_Constitutional_Affairs/Fire_safety/Report/c02

UK Parliamentary Debates, Hansard, Fire Precautions Bill, 20 November 1970, Volume 806, Order for Second Reading.

3 Principles of fire

3.01

The science of fire is a complex discipline. This chapter provides an introduction to the rudiments of the subject to give occupiers a basic knowledge of different types of fire that may happen in the home and how to minimise associated risks. This chapter is not a comprehensive study of fire development and is no substitute for further learning. However, it is important for the reader to understand basic elements of ignition sources, types of fire, formation and spread, and prevention and separation in the home. Together, these elements form a core set of considerations for the task of assessing fire risk.

3.02

It is important for occupiers to have a basic understanding of the likelihood and impact of a fire in their home. They need to keep the risk as low as possible and to make an escape plan. Building owners and managers, however, will need a more comprehensive understanding, especially if they have responsibility for undertaking a suitable and sufficient risk assessment. Just as importantly, building owners and managers need a strong grounding in this area to help them recognise the limits of their own expertise and when to defer to professionals equipped to undertake adequate fire risk assessments. Further reading on this topic area is recommended to increase knowledge.

3.03

Fire may be defined as 'the state of combustion in which inflammable material burns producing heat, flames and often smoke.'[1] Fire is a chemical chain reaction requiring a source of ignition, fuel and an oxidising agent that results in heat and flames (light energy) and, generally, smoke. In the home, the oxidising agent is unlikely to come from any source other than naturally occurring oxygen. So long as one of the elements is available, the chain reaction will continue until it either runs out of fuel or is extinguished to remove heat or oxygen.

3.04

There are several types of ignition to be aware of in the home. These are not always obvious:

- direct sources of open flames, including matches, candles and gas hobs;
- overloaded electrical circuits or defective kitchen appliances in the home, which can be an unexpected and overlooked source of ignition (as was the case in the Grenfell Tower fire);[2]
- radiation; for example, where the heat from a portable electric heater is left too close to flammable items such as curtains or drying clothes. Eventually the ignition temperature will be met, and the fabric will combust. London Fire Brigade figures indicate that 819 fires were caused by electrical heaters in the five years between 2015 and 2020;[3]
- conduction of heat energy transferred from one place to another by direct contact; the probable source of a devastating house fire in Buxton, Derbyshire, in 2018 was an exploding pressurised aerosol canister left on a wood burning stove.[4] Just about every household is likely to have several aerosols canisters at any one time. These should be treated as a serious fire risk and should not be left on a sunny window sill.

3.05

There are multiple sources of fuel in the home that will burn very easily. A quick look around an average home will give an indication of the large number of things that will burn. The range and quantity of flammable domestic goods has increased significantly over the years, notably plastic items and computers, printers and TVs, all of which require a source of electrical power. How safe is your house? How safe are your tenants' homes?

Combustion process

3.06

The speed at which a fire can develop depends on:

- the types and quantity of fuel;
- the type and strength of the ignition source;
- the amount of oxygen available;
- the size of the compartment in which the fire is located.

During the combustion process, available fuel is ignited when the ignition temperature is met and heated to the point at which it eventually releases gases that ignite and flame (the flash point). It is not the organic material that burns (wood, for example) but the vaporised gases that are released from the material. Organic materials in all states of solids, liquids and gases have the ability to ignite, depending on individual characteristics. Thin wooden twigs will ignite and burn much more easily than a larger log. (This is why matches burn readily and are part of everyday life.) Following ignition, the heat of the flame will keep the combustion process sustained as long as the flames are hot enough to remain at the fuel's ignition temperature. The combustion process will end when the fire runs out of fuel or the heat or oxygen is taken away. The removal of heat/oxygen is the principal strategy used in firefighting. Many of us can recall, as children, garden bonfires that created a great deal of smoke before they eventually burst into flames, as the flash point of the garden waste was met. We may also be able to recall similar bonfires that did not burn because they failed to get hot enough to release the volatile gases that create the flames and heat – the ignition source was too weak to vaporise the gases.

3.07

Combustion can either be complete or incomplete, depending on the amount of oxygen available. During complete combustion, a blue flame is produced, which is typically seen on a gas hob. For this to happen, there needs to be enough oxygen to combine completely with the fuel gas. Incomplete combustion occurs wherever there is insufficient oxygen. This produces carbon products such as smoke, soot, carbon monoxide and other poisonous gases, as in a house fire. It is often the case in a house fire that occupiers – particularly if they are asleep – are initially overcome not by the flames but by the smoke and poisonous gases (for example, carbon monoxide, carbon dioxide and hydrogen cyanide).

3.08

The building materials used to construct or refurbish residential buildings are particularly important when assessing the risk of fire and deciding what the appropriate precautions may be. Clearly it would be a great deal harder to ignite and sustain a fire in a concrete and steel building than in a modern timber-framed building.

Types of fire

3.09

Understanding the different types of fire and how they form and behave is important when considering, as part of a thorough risk assessment, the most appropriate passive and active fire precautions necessary to reduce the risk of harm from a fire. Passive precautions include appropriate compartmentation, fire barriers and fire doors. Active fire precautions include smoke and heat alarms, portable extinguishers and sprinklers. Chapter 7 explores these elements in greater detail in practical settings. The following paragraphs provide some basic information about different types of fire.

3.10

In the early stages of a fire rapidly forming in a well-ventilated room, the rolling smoke and hot gases rise quickly as a plume until they reach the ceiling. From there, they spread sideways until they meet a wall and cool slowly before filling the compartment. How quickly the compartment fills depends on the ignition source, fuel and ventilation. There may well be parts of the upper corners that remain smoke-free for a short while. The smoke will then cool and mix with the surrounding air and spread through any gaps in the construction and finally under doors into other parts of the building. The behaviour of the smoke plume is relevant when considering where on a ceiling to affix a smoke alarm – a poorly located alarm may not activate rapidly enough to provide early warning.

3.11

A slow, smouldering fire in the home is typically caused by ignition from a small, weak, direct heat source, for example, a smouldering cigarette that was carelessly disposed of. It will generate increasing quantities of smoke before the combustible gases become hot enough to burst into flames. The Sandy Spring Volunteer Fire Department reported that 'smoking materials started an estimated 17,200 home structure fires reported to U.S. fire departments in 2014 which caused 570 deaths, 1,140 injuries and $426 million in direct property damage.'[5]

3.12

A flashover occurs where the heat energy from a fire is radiated back into a compartment. For example, heat from flaming surfaces (walls and ceilings) raise the temperature of the contents of a room – furniture, carpets and curtains – to their ignition temperatures, so that they all ignite simultaneously. At this stage the fire is fully formed and at its most dangerous. The spontaneous rapid increase in energy and very high temperatures produced may blow out windows as a consequence of pressure changes. Not all fires will result in a flashover. This depends on the size of the space and availability of fuel and oxygen. A small living room fire in a shared house with plenty of combustible items is far more likely to result in a flashover fire than a larger room with higher ceilings and fewer items in it.

3.13

A backdraught can happen in a compartment filled with significant amounts of partly combusted gases that have insufficient oxygen to burst into flames fully. This could be due to limited amounts of ventilation, for example, where a fire door is very tightly fitted to a carpet, preventing adequate airflow under the door. When the door is opened, the oxygen allows the fire to erupt rapidly. This can result in an explosive mix directly onto whoever opened the door.

3.14

A chip pan fire is an example of a fast-flaming fire, which may also produce smoke, depending on how often the oil has been used. The numbers of chip pan fires in dwellings in England are slowly dropping. However, they are still a common occurrence, despite awareness campaigns and the availability of safer, thermostat-controlled deep-fat fryers. In 2018–19, in incidents where the fire and rescue service attended in England, chip pan fires caused around 5 per cent of all accidental primary dwelling fires that resulted in casualties.[6] There are likely to be far more fires than the official statistics suggest, as fires may not be recorded if fewer than five pumping appliances attend. Not only are chip pan fires themselves extremely dangerous, but the situation can be made far worse if an attempt is made to put out the fire using a water extinguisher. The water will vaporise instantly and explode because of the oil's high temperature.

Fire classification

3.15

In the UK there are six different types of fire classified in British Standard BS EN2:1992. These are also used in Australia and New Zealand. Different types of portable fire extinguishers have been designed to tackle specific types of fire within the British Standard taxonomy. Understanding the cause of a fire is important, as using the wrong type of extinguisher could be disastrous.

- Class A – fires involving solid materials such as wood, paper or textiles of an organic nature;
- Class B – fires involving flammable liquids or liquefiable solids such as petrol, diesel or oils;
- Class C – fires involving flammable gases;
- Class D – fires involving flammable metals;
- electrical – fires involving live electrical apparatus or electrically energised equipment (this class has not been designated a letter);
- Class F – fires involving cooking oils and fats such as in deep-fat fryers.

Types of fire alarms

3.16

When deciding what type of fire alarm system to install in a particular building or development, an important consideration is what type of fire is most likely to occur, based on the outcomes of an assessment of risk and dwelling use. If the incorrect type of alarm is used in the wrong location, either it will not sound at all or it will misfire regularly with the increased likelihood of the occupiers eventually ignoring the alarm. Most smoke or heat alarms now combine the detector and sounder functions. However, some properties will still have older systems that consist of a single-function detector linked to a separate bell or klaxon.

3.17

Commonly, alarms provided in the home either respond to smoke or heat. However, these alarms are also available in different types to suit different locations and the protection level they are expected to provide to occupiers. The choice and location of alarm depends on the characteristics of the fire anticipated and what warning is necessary.

3.18

Two types of heat alarm are widely available:

- fixed temperature alarms will respond to an increase in the ambient temperature and sound when the air temperature increases above a set temperature;
- fixed temperature/rate-of-heat-rise heat alarms will not sound where there are slow fluctuations in the temperature, but will sound in the event of a rapid increase in temperature consistent with the outbreak of a fire.

3.19

If a heat alarm is mislocated – such as on a ceiling too close to an oven – there is a strong chance that the alarm will misfire, for instance when an oven door is opened and hot steam rapidly increases in the surrounding air.

3.20

Smoke alarms are usually either ionization or optical alarms:

- an ionization alarm contains two electrodes with a positive and negative charge that create a small electrical current. When smoke enters the chamber, the electrical current becomes unbalanced. The alarm will sound once there is sufficient smoke in the chamber. Ionization alarms are very effective at detecting very small, almost invisible smoke particles given off by fast-flaming, virtually smoke-free fires. They are, however, prone to misfiring from larger smoke particles produced by cooking. For this reason, they should not be located close to kitchens;
- optical alarms work on the principle of smoke disrupting a pulsed infra-red LED source when it enters the alarm chamber. The smoke

scatters the LED light onto a light detector that generates electricity and sounds the alarm. These alarms are more sensitive to denser smoke.

3.21

It is essential that the most suitable type of alarm is chosen for the identified risk. This is particularly important when an alarm needs to be retrofitted to mitigate an increased risk of fire. Equally, alarms must be correctly located in order to ensure their effectiveness and keep accidental misfires to a minimum.

Compartmentation

3.22

Compartmentation is an important part of passive fire protection. The purpose of compartmentation is to stop or slow down the spread of smoke and hot gases between individual units of accommodation through the effective subdivision of the building. This a particularly important and effective strategy in high-rise blocks of flats and bedsit accommodation. All dwellings require some element of fire separation, depending on the design and proposed use. Effective compartmentation is achieved through:

- fire-resistant floors, walls and ceilings;
- properly fitting fire doors;
- cavity barriers;
- fire-stopping through gaps where services run.

3.23

In the construction, conversion or refurbishment process, it is vital that compartments are properly built: all gaps and joints must be adequately filled and sealed using the correct materials in order to ensure that no flames or gases can penetrate. Problems with compartmentation tend to arise at a later stage in the life of a building. In particular, issues are likely to emerge when subsequent conversions, alterations or improvements pierce the original compartment fire barrier without remedy. After the major fire at Lakanal House in 2009, the London Fire Brigade conducted a post-incident inspection that identified pre-existing deficiencies in the block. These included a lack of compartmentation in the suspended ceilings and a lack of fire strips and smoke seals to fire doors. For these failings, Southwark Council was prosecuted and heavily fined in 2017.[7]

Fire doors

3.24

Fire doors are an engineered safety product that, when properly hung and fitted, are one of the most important elements of fire protection in a building. They are designed to resist flames and stop fire spread in order to help protect life and aid effective escape. Fire doors that are fitted into an existing frame in a dwelling cannot be guaranteed to provide 30 minutes' protection. It is only as a set that the door, frame and ironmongery achieve the overall performance, reliability and integrity necessary to provide 30 minutes' protection in the event of a fire. Fire doors require heat expanding intumescent strips and often cold smoke seals to slow the escape of smoke and hot gases.

3.25

According to data from the Fire Door Inspection Scheme in 2019, lack of knowledge about the correct installation and maintenance of fire doors is common.[8] Out of 100,000 doors inspected, 76 per cent were condemned as not fit for purpose. The top three reasons given were excessive gaps of over 3mm, issues around smoke sealing and poorly adjusted closers. From more than 2,700 buildings inspected, 30 per cent of doors were condemned for poor installation.

3.26

Experience of inspecting HMOs and purpose-built student accommodation can readily back up these findings of common problems with installation and ongoing management. If a specialist gap tester is not on hand, one-pound coins can serve as a useful substitute for measuring the adequacy of a fire door's fit within its frame. A one-pound coin is 2.8mm thick, so if two coins fit into a gap between a door and its frame, the gap is excessive (see Image 3.1) and will allow the escape of combustion products.

3.27

Fire doors are heavy, so the hinges must be fitted with the correct number of screws to retain integrity and prevent the door from falling in its frame, something that will prevent the door from closing properly (see Image 3.2).

Image 3.1 Excessive gap between the door and frame of a newly hung fire door in a newly built student accommodation block where building control completion certificate was issued. Para 3.26

Image 3.2 Screws missing from the hinges of a fire door in a newly built student accommodation block where building control completion certificate was issued. Para 3.27

Fire-stopping around pipes

3.28

Effective fire-stopping is critical, especially around pipes. To support compartmentation, larger holes, gaps and joints should be sealed effectively. This includes any gaps around pipes where they penetrate walls or other barriers. Intumescent collars or properly cut barrier materials should be used for this purpose in order to prevent any leakage of smoke and gases (see Image 3.3).

3.29

Pink polyurethane expanding foam is a commonly used product for filling gaps. It is, however, often used incorrectly. Where it has been misused, it is unlikely to meet Building Regulations. It is therefore important, when using PU foam fillers or when considering their use, to follow the manufacturer's instructions and generally to restrict their use to linear gaps between 10 and 30mm wide.[9]

Image 3.3 Correct fire stopping around pipes through fire partitioning that had been certificated by the installer to meet the fire separation standard. Para 3.28

Notes

1 Collins English Dictionary, Millennium Edition, 1999
2 Dr J. Duncan Glover, *Expert Report for the Grenfell Tower Inquiry*, 15 October 2018, pp. 6–7
3 London Fire Brigade, *Portable Heaters, Gas Fires and Open Fires*, www.london-fire.gov.uk/safety/the-home/portable-heaters-gas-fires-and-open-fires/
4 Derbyshire Fire and Rescue Service/Derbyshire Fire and Rescue Authority, press release, *Firefighters Issue Warning After Aerosol Can Causes Explosion In House In Buxton, Derbyshire*, 30 April 2018, www.derbys-fire.gov.uk/application/files/1115/6077/6251/PR_AshStreetBuxton_April_18.pdf
5 Sandy Spring Volunteer Fire Department, *Smoking Fire Safety*, www.ssvfd.org/safety/smoking-fire-safety/#:~:text=Smoking%20materials%2C%20including%20cigarettes%2C%20pipes,U.S.%20fire%20departments%20in%202014.&text=Smoking%20materials%20caused%205%25%20of,property%20damage%20from%20home%20fires
6 See Home Office fire statistic table 0601, *Primary Fires in Dwellings and other Buildings, by Cause of Fire, England – Fire and Rescue Incident Statistics: England, Year Ending December 2019, updated May 2020*
7 See Inside Housing, *Council Fined £570,000 for Lakanal House Fire Failures*, 28 February 2017
8 See Fire Door Inspection Scheme, https://fdis.co.uk/
9 See LABC, *Beware of PU Foam Fillers*, 30 April 2015, www.labc.co.uk/news/beware-pu-foam-fillers

Further reading

Department for Communities and Local Government/Chief Fire and Rescue Adviser, *Fire and Rescue Service: Operational Guidance: Generic Risk Assessments: Flashover Backdraught and Fire Gas Ignitions*, August 2009, www.ukfrs.com/sites/default/files/2017-09/GRA%205.8%20Flashover_%20backdraught%20and%20fire%20gas%20ignitions.pdf

4 Fire safety risk assessment and fire protection in residential premises

4.01

Fire risk assessments (FRAs) were introduced by the Regulatory Reform (Fire Safety) Order 2005 (FSO), which required a responsible person to undertake an assessment of risks rather than wait to be directed by the fire service to impose fire safety measures.[1] In England and Wales, an FRA is required for non-domestic premises, including hotels, guest houses, the common parts of blocks of flats, bedsits and student accommodation blocks. In England and Wales, the responsible person must undertake a 'suitable and sufficient assessment' of the fire risks to 'relevant persons' on the premises and also in the immediate vicinity. Alternatively, they can engage someone to complete the assessment on their behalf. In Scotland[2] and Northern Ireland,[3] this responsibility is borne by the 'duty holder.'

4.02

A suitable and sufficient assessment must address the 'general fire precautions.'[4] These include:

- fire safety management arrangements;
- testing and servicing the fire alarm;
- keeping records of evacuation drills if necessary;
- ensuring the means of escape can be effectively used;
- providing information for occupiers and fire authorities;
- keeping and updating an action plan of fire prevention and control procedures necessary for the building.

The fire and rescue authorities are charged with enforcing the regulations, and they bear the responsibility of having to determine whether or not an assessment is suitable and sufficient for the premises. They have a range of powers available to them to force improvements and to prohibit the use of dwellings for sleeping.

4.03

Not all sleeping accommodation is required to have an FRA under the FSO, for example, single household dwellings, shared houses, cluster flats and studios in purpose-built student accommodation (PBSA) blocks, and individual flats within blocks. These buildings fall within local authority regulation in England and Wales under the Housing Act 2004, Part 1 (HHSRS). The regulatory scope includes the hazard of fire, allowing officers to assess the risk of exposure to uncontrolled fire and smoke. Powers to require improvements are vested in authorised LHA officers to reduce identified risks if necessary. They can, for example, require additional fire precautions to be installed.

4.04

To ensure that all sleeping accommodation can be effectively regulated, there is an overlap in England and Wales between the FSO, the Housing Act 2004 and also, in England, the Smoke and Carbon Monoxide Regulations 2015. There is a dual enforcement regime, operated between LHAs and fire and rescue services. Meanwhile, in Scotland, all sleeping accommodation is regulated by the fire authorities.

4.05

In England and Wales, there is no requirement for the responsible person or fire risk assessor to have any particular fire risk qualifications or to be competent to undertake an assessment. Currently, anyone can set themselves up in business as a fire risk assessor, because it remains an unregulated activity. This is not accidental: it allows an owner of the very smallest and simplest of dwellings to avoid having to engage a specialist consultant, as they (the owner) are the responsible person. The landlord of a small, two-storey shared house or small hostel, who is the residential manager and responsible person, is likely to be capable of identifying the fire risks and fully understands the management and necessary fire precautions to undertake a suitable and sufficient assessment. However, in a large block of flats or student accommodation, the landlord may have little control over how the building is operated, having engaged an operational management company. In this situation, the responsible person may be a senior executive in the management company with little or no knowledge of the day-to-day management of the block. But they still have responsibility for the assessment, so it is reasonable to engage an independent fire risk assessor. The responsible person must either demonstrate that they have sufficient knowledge and the requisite skills or engage someone who has.

4.06

An FRA is not a tick-box exercise. It is a thorough process for assessing the hazard and then managing and reducing the risks identified. Effective risk assessment is an ongoing process that requires constant review to ensure it remains suitable and sufficient. Landlords and building managers cannot always be on their premises, but they must ensure that the occupiers are safe and know what to do in the event of a fire.

Whoever completes the fire risk assessment must have sufficient and necessary skills and awareness that they are able to:

- identify the fire hazards;
- identify relevant people at risk, particularly sleeping occupiers who may be less familiar with the building, and people with disabilities and vulnerabilities. A personal emergency evacuation plan (PEEP) should be drawn up, where necessary. This is a bespoke emergency escape plan required for individuals who will need assistance in evacuating a building to an ultimate place of safety;
- understand and identify the necessary fire safety measures to protect people, including management of escape routes and fire management arrangements;
- review the fire safety management arrangements (for example, routine inspections of the escape routes, fire drills, test inspection and maintenance of fire precautions);
- record the significant findings and prepare an action plan;
- effectively communicate the significant findings of the assessment with all relevant people;
- keep the assessment under regular review to make sure it remains relevant and current.

4.07

The responsible person should be capable of identifying when an FRA is required, and that person should understand how it links with the construction and occupation of the elements of a building. Buildings must be properly designed and built by skilled professionals who are competent, for example, at fitting a fire door or installing a fire alarm correctly. Ideally, an FRA should initially be undertaken after a building or conversion has been completed and at first occupation – and then kept under constant review. Ideally, FRAs should not be undertaken before the completion of the building and used by the landlord as a snagging exercise to identify poor construction, for example, badly fitted fire doors or gaps and holes in compartmentation where pipes have been run and not properly sealed. However, this is often the case.

4.08

It is difficult to undertake a fire safety risk assessment or review in a building without having conversations with the managers, employees or contractors (for example, night-time reception or security) and without thorough scrutiny of management fire safety documents. The importance of management conversations to support risk assessment cannot be overstated.

Judging the competence of a fire risk assessor

4.09

Irrespective of who undertakes the risk assessment, the responsible person must ensure the assessment is suitable and sufficient, as the duty cannot be delegated. Where a responsible person engages a consultant, they should assure themselves that the assessor is competent. In December 2011, the Fire Risk Assessment Competency Council agreed to industry-wide criteria to help responsible persons judge competence. A person is regarded as competent when they have sufficient training and experience or knowledge and other qualities to enable them to prepare a suitable and sufficient risk assessment.[5] A competent assessor will understand and acknowledge the limits of their expertise. Not all risk assessors may be sufficiently knowledgeable or experienced to assess a very tall building. If the responsible person engages a third-party assessor, they should undertake appropriate checks to make sure the assessor is competent for the task. Examples of competence checks include:

- requesting evidence of training certificates or registration with a third-party accreditation scheme;
- asking for examples of other, similar premises where the prospective appointee(s) has undertaken assessments and getting references from the duty holders;
- drawing up a clear scope of work and obtaining more than one quotation to allow for thorough comparisons;

- securing proof that the assessor(s) has sufficient professional indemnity insurance to undertake the work;
- retaining all written records to lay an audit trail of the steps taken when selecting an assessor.

Because the legislation varies slightly across the UK constituent nations, it is important that an assessor has a clear understanding of the specific legal requirements and guidance that apply to the country in which the assessment is taking place.

4.10

In complex buildings, an assessor must fully understand not only the general construction and layout of the building and the passive and active fire protection systems provided, but also the standard of management. As an integral part of their work, assessors should produce a thorough action plan highlighting significant findings of both the building and the fire safety management systems. This should be compiled objectively in order to guarantee the integrity of the assessment's outputs, even at the risk of losing a customer who dislikes the assessor's findings and conclusions.

4.11

Failure to engage a competent assessor can have serious consequences for the responsible person. If, as a result of a failure to undertake a suitable and sufficient risk assessment, someone is put at risk of serious injury or death, the responsible person has committed an offence that can lead to a heavy fine or, in extreme cases, imprisonment. In 2017, a consultant risk assessor in England was handed a suspended prison sentence for producing a 'woefully inadequate' fire risk assessment for a nursing home. The company that commissioned the assessment (responsible person) was fined £40,000 plus £15,000 in costs for four FSO breaches. The prosecutors successfully argued that the responsible person had left elderly clients at risk of death or serious injury in the event of a blaze.[6]

Types of fire risk assessment

4.12

There are four types of fire risk assessment. Responsible persons should ensure that they are commissioning the correct type and that they are not influenced by cost. If they cut corners, the assessment type chosen may not extend to accessing all relevant parts of a building. A competent assessor may well recommend that an elevated type of assessment is appropriate, but, in the end, it is up to the responsible person to decide what is necessary. The types of fire risk assessment are as follows:

- Type 1: common parts only (non-destructive). This is the basic-level assessment that will satisfy the requirement of the FSO. Generally, a Type 1 FRA in a small building (for example, a hostel or small, low-rise block of flats) will be adequate;
- Type 2: common parts only (destructive). This is similar to Type 1 but includes opening up construction, which must be made good to the original fire resistance standard on completion. Type 2 assessments are generally undertaken if there is suspicion that building deficiencies would result in a fire spreading beyond a flat or bedsit of origin – for example, evidence of missing cavity barriers or an inspection above a suspended ceiling in a converted building;
- Type 3: common parts and flats (non-destructive). These risk assessments go beyond the scope of the FSO but are appropriate under the Housing Act 2004 – for example, when thoroughly assessing cluster flats in student blocks of accommodation. This type of assessment may be appropriate in older blocks of rented flats or HMOs where there is evidence of alterations, damage to the fire separation between the flats or a poor standard of management or maintenance of flat fire doors;
- Type 4: common parts and flats (destructive). This is the most comprehensive type of risk assessment that could, for example, be relevant on sale of a large block of flats with clear alterations undertaken or a poor repair record.

4.13

Risk assessments must be reviewed regularly and kept up to date. The FSO does not prescribe a time period between assessments. Typically, an annual review is deemed sufficient. However, in reality, a basic review is often undertaken every time the managers or representatives visit the building – for example, to verify that the escape route is clear.

Assessment risk in practice

4.14

The importance of a suitable and sufficient fire risk assessment, effective and knowledgeable management and robust regulation cannot be overstated. Where poor building management is identified, well-trained and knowledgeable officers who can judge the fire risk must be relied on to provide a safety net for occupiers. Officers must have an adequate understanding of building elements and building failure to decide and force improvements to minimise the spread of products of combustion, for example, justifying what type of early warning systems are appropriate in individual dwellings.

4.15

Only the fire and rescue services can judge whether an FRA is suitable and sufficient. However, the dual regulation regime in England and Wales means that inspections of residential premises are made by either the fire service or LHA officers, depending on locally agreed arrangements. As FRAs are not always required in smaller dwellings regulated by LHAs, it is important that officers who make fire safety regulation decisions have sufficient understanding of the actual fire risks that they are seeking to reduce. These include:

- possible ignition sources;
- building and compartment performance and possible failures;
- fire growth and spread;
- potential occupier behaviour.

Once they have achieved a proper understanding of the risks, officers can begin to consider what additional passive and active fire precautions will be required to reduce those risks. Officers must base regulatory decisions – including service of enforcement notices – on reducing the actual risk. What they should not do is simply follow what a guidance document states for a particular type of dwelling without due regard to the specific issues and risks of the particular building under assessment. Officers must have sufficient knowledge and understanding of fire risk to make a judgement to take appropriate action and also to defend the service of a notice or order in a Tribunal.

4.16

Fire risks can vary according to how a dwelling is occupied and managed. One year, a house may be responsibly managed and occupied by a group of non-smoking young professionals who all know each other. The fire risk is likely to be low, similar to that of a single family. If the same house is then sold and rented at a later stage to more vulnerable individuals who smoke and have little or no social cohesion, and the new owners are poor managers, then inevitably the fire risk will increase. The house has not changed, but the occupation and management have.

4.17

Each dwelling is different, so it is important to consider as part of a risk assessment:

- the number of storeys;
- overall floor area;
- layout;
- use of rooms (particularly if they are inner rooms used as sleeping rooms located off a kitchen or living room);
- travel distances to a place of safety.

Where the landlord of a smaller building (for example, a house or flat in multiple occupation) undertakes the assessment alone, the following should be considered. Table 4.1 sets out a typical range of risks commensurate with domestic living, but it should be noted that this schedule is not exhaustive.

Table 4.1 Typical fire risks associated with domestic living

Occupiers and other relevant persons	**Sources of ignition**
• number of sleeping occupiers permitted • location of sleeping rooms – consider inner rooms located off kitchen or living room • presence of occupiers with disabilities that impede/prevent rapid escape • numbers and ages of children • management team and contractors • smoking, alcohol and drug-taking that will impair hearing and compromise response to an alarm	• smoking and candles and other naked flames • electrical items, including portable appliances and overloaded electrical extension boards (particularly in bedrooms); all small appliances should be CE-marked • unattended cooking • use of portable heaters
Sources of fuel	**Fire protection measures**
• build-up of unwanted mail, circulars and newspapers • cooking oils and other flammable liquids • clothing, textiles and bedding • furniture • gas supply • greasy cooking appliances, grill pan and extractor hood	• Is the grade of fire alarm adequate and sufficiently loud to rouse all sleeping occupiers? • Are there gaps and holes in the compartmentation that require sealing – particularly in cupboards and voids? • Is the standard of fire separation between rooms adequate? What about glazed transoms over doors?

- What is risk of rapid spread of flames in the escape route, for example, as a result of highly-varnished internal decorative wooden lining?
- Is emergency lighting needed? This depends on the length and complexity of the escape route and provision of windows in the escape route.
- Where portable fire extinguishers or a blanket are provided, have the occupiers received adequate instruction?
- Are there secondary emergency escape windows, if necessary?

Regular fire safety management

- regular inspection of fire doors and fire alarm
- replacement of missing and defective intumescent strips and cold smoke seals
- arrangement to keep the escape route clear
- keeping of records in log books

Fire safety information

- Can all the occupiers understand the fire safety information, for example, if their ability to understand or read English is poor?
- Has information been given to tenants about what to do if the alarm rings?
- Should tenants conduct a full evacuation or stay put?
- Who summons the fire and rescue service?

Enforcement of FSO

4.18

Each fire and rescue authority has to decide how to enforce the FSO, which can include inspections of residential dwellings. The inspection regime is generally risk-based and is focused on larger, higher-risk buildings that have more occupiers (particularly where the occupier profile includes vulnerable or elderly people, such as in care homes). The authorities have a range of powers of entry and enforcement powers, where enforcement action is considered (for example, as a result of a complaint or following a fire).

4.19

If an authority determines that a set of premises constitutes a serious fire risk, they may issue an alterations notice, which prohibits any change to the premises without formal consultation with the authority.[7] While the notice is in force, the responsible person must advise the authority of any changes that may increase the fire risk. They must also submit a copy of the fire risk assessment with the proposed changes to the fire safety arrangement.

4.20

An authority can serve an enforcement notice when it considers the FSO has not been complied with.[8] A notice will set out points of non-compliance and remedial actions to be taken within a specified timescale. Failure to comply with this type of notice is an offence that can result in prosecution.

4.21

A prohibition notice can be served in extreme cases where the authority considers that the risk to life safety is so serious that it can restrict or close the premises completely.[9] The Greater Manchester Fire and Rescue Service served a prohibition notice on a six-storey, ten-year-old block of flats in Oldham in October 2020 because of serious defects that would affect means of escape.[10] Where the authority serves a notice on an HMO, they must notify the LHA, where practicable.[11] However, there is no duty placed on the authority to rehouse displaced occupiers following the service of a notice. This means that tenants of private rented dwellings may be made homeless. The notice remains in force until the authority is satisfied that the premises are safe.

Impact of Building Regulations on fire risk

4.22

All residential buildings should be built or converted to protect occupiers from the risk of fire to meet the current Building Regulations. In recent years the building regulation processes have fundamentally changed following deregulation.

4.23

Since 1 January 2001, builders have had a choice of engaging a private approved inspector or the local authority building control to have plans and building works inspected to meet the building regulations.[12] To ensure the impartiality of approved inspectors, Regulation 9 provides that 'approved inspectors shall have no professional or financial interest in the work they supervise unless it is minor work.' The UK Government has accepted that the building regulation system is not perfect, and the construction industry has ultimately to take responsibility for delivering safe buildings:[13]

> *Competition between local authorities and approved inspectors in the provision of building control services throughout England & Wales can provide a stimulus to greater efficiency and higher standards of service to the customer. However, these same market forces have the potential to drive down building control standards which could put at risk the health and safety of building users. . . . Building control does not remove the obligation of the person carrying out the work to achieve compliance with the Building Regulations, so the construction industry has an enormous part to play.*

4.24

Where it can be established that building works undertaken do not meet the Building Regulations, the local authority Building Control departments have a range of enforcement powers to resolve the non-compliance. A local authority may prosecute a person (for example, the builder or main contractor) for undertaking building work that contravenes the Building Regulations (s.35 and 35A of the Building Act 1984). Prosecution is possible up to two years after the completion of building work. Under s.36 of the Building Act 1984, an enforcement notice requiring the alteration or removal of non-compliant works must be served within 12 months of the date of completion of the offending works. It is likely that these offences may only come to light through a complaint, possibly made by a neighbour.

4.25

The enforcement window is therefore relatively narrow. As a consequence, the job of identifying and improving poor fire safety works that put occupiers at risk through poor building often falls to the other regulators, specifically fire and rescue services or LHA officers using powers under the Housing Act 2004.

4.26

The Building Regulations do not impose any requirements on the management of a building but do assume that it will be properly managed. Once the building is in use, the management regime should be maintained and a suitable and sufficient risk assessment undertaken for any variation in that regime.

4.27

Recent fires in blocks of flats show that external fire spread can be rapid and devastating. It is generally human behaviour or unforeseeable electrical failure that leads to ignition. The initial ignition of the Grenfell Tower fire was probably from a fridge/freezer.[14] Occupiers in furnished rented properties may not be aware of electrical appliances that are at risk of ignition, for example, the general recall of Whirlpool appliances in 2019 and 2020 to remedy fire safety faults affecting washing machines and dryers. The fire at the Lacrosse Tower, Melbourne, was probably started by smoking materials.[15]

Notes

1. SI 1541
2. In Scotland, requirements on general fire safety are covered in Part 3 of the Fire (Scotland) Act 2005, supported by the Fire Safety (Scotland) Regulations 2006
3. The Fire and Rescue Services (Northern Ireland) Order 2006 and the Fire Safety Regulations (Northern Ireland) 2010 came into effect on 15 November 2010
4. The Fire Safety Order 2005 Article 4, *Meaning of 'General Fire Precautions'*
5. Fire Risk Assessment Competency Council, *Competency Criteria for Fire Risk Assessors*, Version 1, published 21 December 2011; Version 2, October 2014
6. Fire Safety Law, *Consultant Fire Risk Assessor Receives Prison Sentence for 'Woefully Inadequate' Fire Risk Assessment*, 5 January 2017, www.firesafetylaw.co.uk/consultant-risk-assessor-receives-prison-sentence-for-woefully-inadequate-fire-risk-assessment/
7. The Fire Safety Order 2005, Article 29
8. *Ibid*, Article 30
9. *Ibid*, Article 31
10. See Fire Protection Association, *Prohibition Notice for Oldham Apartment Block*, 19 October 2020, www.thefpa.co.uk/news/prohibition-notice-for-oldham-apartment-block
11. Fire Safety Order 2005, Article 31(6)
12. The Building (Approved Inspectors etc.) Regulations 2000, SI 2532
13. Department for Communities and Local Government Building Control Performance Standards Advisory Group, January 2017, https://assets.publishing.service.gov.uk/government/uploads/system/uploads/attachment_data/file/585965/Building_Control_Performance_Standards_2017_Final.pdf
14. Grenfell Tower Inquiry, Dr J. Duncan Glover's expert report, 15 October 2018, para 16.2(1), p. 77
15. Post-incident analysis report, Lacrosse Docklands, 673–675 La Trobe Street, Docklands, Melbourne, Australia. Metropolitan Fire and Emergency Services Board, 25 November 2014

Further reading

Colin Todd (C.S. Todd and Associates), Report for the Grenfell Tower Inquiry, *Legislation, Guidance and Enforcing Authorities Relevant to Fire Safety Matters at Grenfell Tower*, March 2018, https://assets.grenfelltowerinquiry.org.uk/documents/Colin%20Todd%20report.pdf

5 Essentials of fire safety management in residential dwellings

5.01

The effective management of fire safety in all dwelling types, across all tenures – particularly in high-rise residential buildings (HRRB) and student accommodation – is essential if the risk of uncontrolled fire is to be kept as low as possible. A fire risk assessment must identify the building management process for the responsible person to follow and identify any required improvements in the action plan. The assessment must be prepared and set out in a manner that can be clearly understood by the responsible person, because how building managers respond to a fire risk assessment is particularly important to the ongoing management of fire safety.

5.02

Effective management encompasses inspection, servicing and the maintenance of records for all fire precautions in a building. Precautions include alarms, emergency lighting, sprinklers, the regular inspection of fire doors and making good any damage to compartmentation. Arrangements must be clearly communicated to residents, so that they know how to respond to an incident and how to evacuate the building if necessary. Where highlighted by the fire risk assessment, evacuation drills must be tested to make sure they are effective and appropriate for the building risk. Well-organised and regular inspections of fire precautions form the basis of a sound approach to managing fire safety.

Arrangements for emergency evacuation

5.03

Emergency evacuation arrangements are a fundamental management responsibility. They must include:

- clear lines of communication to residents and other relevant people in the building;
- clear advice on:
  - whether to stay put or evacuate;
  - the importance of keeping escape routes clear;
  - raising the alarm;
  - who should contact the fire and rescue service.

5.04

Arrangements must not rely on the attendance of the fire and rescue service to rescue occupiers, especially from HHRBs. Even in the smallest shared house, occupiers must fully understand what to do in the event of a fire. The responsible person should give the occupiers clear instructions on whether they are expected to tackle a fire or just leave the building and call the fire and rescue service. (This is especially important where portable fire extinguishers or a fire blanket have been provided.) This information should be displayed on the inside of a flat or student room and in a resident's handbook, as well as in the common areas of the escape routes. Managers need to consider how best to explain the escape arrangements to residents whose first language is not English. For taller and more complex buildings, the risk assessment should identify the most appropriate fire emergency plan for the building:

- stay put;
- phased evacuation; or
- simultaneous evacuation.

The responsibility for creating an effective and well-managed evacuation policy lies with the responsible person (in the majority of buildings, the manager). Occupiers cannot just be left to work it out for themselves, as in larger buildings ineffective communication may lead to confusion and delay.

Evacuation strategies

5.05

The concept of a stay-put policy in HRRB blocks was first developed in the UK in 1962.[1] This remains the strategy in the majority of purpose-built flats. It relies on each flat compartment being adequate to resist fire for 60 minutes. It assumes that the residents are safer remaining in their own unit while the fire in a single flat elsewhere in the building either burns itself out or is tackled by the fire and rescue service without it spreading to the whole building. The occupiers of the flat where the fire starts would be expected to escape, raising the alarm after closing the door behind them. (This is why flats have functioning self-closing fire doors.) The building must be managed well, and the escape routes kept clear at all times. Although generally a fire does not spread beyond the flat of origin, there have been important instances where this has not been the case. For example, in the Lakanal House fire in 2009, the stay-put policy was invalidated by breaches in fire separation and by poor building management. The fire spread externally and internally as a result of a compartmentation failure following refurbishment work. The fire killed six people, including three children. All of them had been told by the fire service control to stay put, even when smoke and flames were entering their flats.[2] Where a stay-put policy is palpably not working – for example, where a building is not performing as expected and is in fact quickly failing – the fire and rescue service incident commander may change the strategy to full evacuation during the course of a fire in order to protect lives. This is what occurred in the fire at the Cube, Bolton, in 2019.[3] When they move in, residents should be made aware that, where a fire breaks out, a standing instruction to stay put may be countermanded if unfolding events dictate that evacuation is the better option.

5.06

Stay put is not internationally accepted. For example, in Australia most high-rise buildings are evacuated during fires, and the installation and maintenance of alarm systems is a strict legal requirement. In 2014 in the Lacrosse Tower fire, Melbourne, all occupiers evacuated safely.

Simultaneous evacuation

5.07

Simultaneous evacuation is where all occupiers evacuate a building at the same time to a designated place of safety. It is often used when the standard of compartmentation cannot be guaranteed or the standards of construction may not be adequate, for example, if the standard of external cladding is unknown. Where a strategy of simultaneous evacuation has been determined for a building, an adequate central alarm system is essential. It is important that the sound of the central alarm can be easily distinguished from an individual flat alarm in order to prevent confusion that could lead occupiers to ignore it.

5.08

Following the Grenfell Tower fire, it quickly emerged that a significant number of buildings had been fitted with the same or similar ACM cladding, which had been a critical factor in the spread of the blaze at Grenfell. It was clear that continuing to attach a stay-put evacuation policy to these other buildings would place residents at heightened risk and that safer, short-term solutions were necessary to better protect occupiers until effective lasting solutions could be implemented. One such interim approach is to operate a strategy of simultaneous evacuation, supported by waking watches.[4]

Phased evacuation

5.09

A potential weak point in simultaneous evacuation strategies is that they might create bottlenecks of occupiers on escape staircases. These have the potential not only to produce panic but also to cause delays in the fire service climbing the stairs to tackle the fire. To avoid this, it may be appropriate to adopt a phased evacuation strategy, where the occupiers most affected by the fire are the first to be evacuated, followed by the occupiers of the storeys above and below. In order to be effective, a strategy based on phased evacuation needs to be supported by an advanced, intelligent or addressable alarm system that should be well managed and tested.

Testing and maintenance of fire alarms

5.10

The responsible person must understand how the fire alarm operates and what routine testing, and maintenance is required. Poorly managed and maintained alarms may quickly become ineffective. In HMOs, regular testing and maintenance of the alarms are a legal management duty.[5] In other rented accommodation, the landlord is responsible for testing the alarm at the beginning of each new tenancy.[6] Integrated panel-controlled systems can provide digital information or printouts, including location of a fault or misfire information. The responsible person should record this information accurately in a log book and make appropriate responses. It is poor management practice to ignore fault indicators or regular misfires, as these need to be investigated to prevent recurrences and to make alterations. Tenants may ignore regular false alarms. Worse still, they may cover alarm units (see Image 5.1), which can lead to delay in evacuation. Remedial action may be as simple as removing a sock from a detector where a tenant has been smoking, but frequent false alarms may be attributable to a heat alarm having been poorly located, perhaps in a kitchen. In Image 5.1, the heat alarm has been poorly located too close to the oven. So the occupiers have disabled it to prevent unwanted sounding. The problem can be solved by relocating the heat alarm elsewhere on the ceiling.

The responsible person should seek to minimise false alarms as far as possible by actively managing buildings under their control – including the residents in them. The Fire Industry Association has produced a guidance document to help reduce false alarms.[7]

Image 5.1 Poorly located heat alarm in a shared house kitchen covered by the occupiers following regular misfiring. Para 5.10

5.11

Responsible persons need to show persistence in trying to identify the causes of false alarms. Dialogue with occupiers and thorough investigation are often key to establishing the reasons for false alarms. Providing occupiers with additional education or making a small alteration to alarms may reduce misfires and the chances of an alarm being ignored. In the aftermath of the major fire at the Cube, Bolton, which resulted in the destruction of the six-storey building, students explained why they ignored the fire alarm:

> *The fire alarm was going off but nobody was paying any attention. It goes off all the time, maybe every hour during the day because someone has done something in the kitchen and it's set off the alarm.*[8]

5.12

All grades of alarm require periodic testing. The requisite testing frequency – either weekly or monthly – depends on the grade and complexity of the alarm. Generally, the manager or the occupier can test the alarm, as no specialist knowledge is required. What is important is the making and keeping of adequate records. These should include the date and time that tests are undertaken and verification that all alarms ring where they are interlinked. Regular testing will indicate that the alarms all sound as necessary, but it will not necessarily indicate that alarms respond to heat or smoke.

5.13

All domestic alarm systems must be inspected and serviced regularly to ensure ongoing reliability in accordance with the recommendations of BS 5839–6[9] or similar national arrangements. Panel-controlled alarm systems should be inspected and serviced every six months. Inspection should be undertaken by a competent person, for example, a specialist alarm engineer who understands the installation and what the alarm is supposed to do. It is recommended that interlinked alarms are serviced annually.

5.14

Responsible persons should keep records relating to alarm installation, commissioning, testing and maintenance. As well as being good practice in itself, this discipline serves to lay a demonstrable audit trail of alarm management that will provide important documentary evidence in the event of a regulatory inspection or investigation. Certification of the commissioning of an alarm system may be required by the appropriate enforcing authorities.

Management of escape routes from individual units of accommodation in HMOs

5.15

The primary escape route should be kept as clear as possible at all times. Managers must ensure that passages are kept free of:

- obstructions such as bicycles parked in hallways;
- tripping hazards such as threadbare carpets and internet cables running across the hallway or down the stairs;
- ignition risks such as refrigerators and washing machines left in the escape route.

In discharging this responsibility, managers should rely on an approach that combines:

- frequent management inspections balanced against the occupiers right of quite enjoyment;
- clear and, if necessary, repeated reminders to occupiers that they have a responsibility to keep the escape route clear and that the consequences of leaving it partly blocked might be serious.

5.16

Routine inspections may identify increased ignition risks such as overloaded extension socket boards, evidence of smoking, candle use or barbeques located on wooden balconies. In circumstances where heightened risk has been identified, it may be appropriate to increase the regularity of management inspections to reinforce the fire safety message across the occupier group.

5.17

Where provided, fire extinguishers must be regularly inspected, tested and properly maintained as part of the regular management regime.

Electrical safety

5.18

Across England, records show that the number of accidental fires associated with faulty or misused electrical appliances is falling each year. However, in 2019, there were still nearly 13,000 primary fires.[10] Of these, over 7,000 were caused by faulty/misused electrical distribution equipment.[11] HMO managers must ensure that there are sufficient electrical plug sockets available in shared kitchens and bedsits, and they must advise occupiers not to overload extension socket boards in the bedrooms. These injunctions hold for all householders across all tenures:

- do not overload sockets with appliances that draw high electrical currents, for example kettles, toasters and hair dryers;
- keep portable heaters away from sources of ignition (for example, curtains) and never cover them (for example, with clothes);
- small electrical devices (for example, phones and tablets) should only ever be charged on hard surfaces (for example, a desk). Occupiers should switch off and unplug chargers if they are going to be away from the property for an extended period.

Electrical Safety First has created a socket calculator that gives useful advice on how to help prevent overloading sockets and the risk of electrical fire.[12]

5.19

Since 2016, electrical consumer units installed in new residential accommodation have been enclosed in non-combustible materials made, for example, out of metal.[13] However, older consumer units are typically enclosed in moulded thermoplastic, which offers no fire resistance. The Electrical Safety in the Private Rented Sector (England) Regulations 2020 came into force on 1 June 2020. The primary objective of the regulations is to reduce the number of electrical fires in the rented sector. Other arrangements apply in Scotland and Wales.[14] Up until 2020, there was no legal requirement for a landlord, unless managing an HMO, to obtain an Electrical Installation Condition Report (EICR). LHA officers had to prove disrepair using the HHSRS and take formal action. This was often difficult because the officer had to have reasonable suspicion that the installation was unsafe and engage a competent person to undertake a test to which a cost was attached.

5.20

The regulations apply to new tenancies from 1 July 2020 and existing tenancies from 1 April 2021. Guidance has been published to help landlords, tenants and local authorities (non-statutory) to understand their roles and discharge their responsibilities. The regulations require landlords to have the fixed electrical installation tested by a competent person no less frequently than every five years and to obtain an EICR. The report will record whether the installation is satisfactory or unsatisfactory and set out any further investigative or remedial works needed to improve safety. These the landlord must complete within 28 days. Some landlords provide occupiers with portable appliances, that is, any appliance with a plug attached (for example, white goods, toasters and lamps). In England, if a landlord provides portable appliances, there is currently no legal requirement for them to get such equipment tested, but it is strongly recommended that portable appliance tests (PATs) are undertaken.

5.21

Risk assessments are unlikely to extend to white goods such as fridges, freezers and dryers, although the likely source of ignition of the Grenfell Tower fire was a faulty kitchen appliance (see Para 3.04). White goods can present a fire risk, and widely publicised product recalls across the world attest to how common a problem this is. Product safety recall lists are published on the Register My Appliance website, along with advice to purchasers on how to register appliance owners with the manufacturer so that they can be contacted if a fire risk is identified.[15]

Portable fire extinguishers

5.22

There is no doubt that the correct use of portable fire extinguishers can help put out a small fire in the early stages and help rapid escape. However, the incorrect use of a portable extinguisher can make a fire worse or injure or damage the health of an untrained user. Where portable extinguishers are installed following a risk assessment, landlords must give sufficient instructions if they expect tenants to use them. Powder extinguishers are no longer advised for indoor use, as the sudden discharge of powder can fill the escape route, affect vision and breathing and slow down escape.[16] Where provided, the responsible or other competent person should undertake an inspection each month, principally to ensure

Image 5.2 Portable fire extinguisher used to prop open a fire door in a shared HMO. Para 5.22

that the extinguishers are in the proper location, have not been discharged or lost pressure and are not, for example, being used to prop open a kitchen fire door (see Image 5.2).

Management and maintenance of fire doors

5.23

Fire doors need regular inspection and maintenance to ensure they remain effective:

- the fire door self-closing devices should be checked at each management visit to ensure they close or latch fully into the frame. The devices may need regular adjustment, especially enclosed frame chain closers;
- tenants must be advised not to fit over-door coat hangers to fire doors, as the extra weight can prevent the self-closing device from latching the door unaided and put strain on the hinges;

- any missing intumescent strip or cold smoke seal must be replaced (see Image 5.3);
- when door frames are painted, the cold smoke seal must be protected from paint damage (see Image 5.4). Fire doors must be routinely inspected to make sure the fire strips have not fallen out of the grooves or become paint-damaged, as such disrepair makes the smoke seals ineffective.

Image 5.3 Missing length of combined intumescent strip or cold smoke seal from a fire door in a student HMO. Para 5.23

Image 5.4 Paint-damaged smoke seal in a fire door. Para 5.23

Notes

1. See Inside Housing, *Where Did the Stay Put Policy Come From and Where Do We Go Now?* 31 October 2019, www.insidehousing.co.uk/insight/insight/where-did-the-stay-put-policy-come-from-and-where-do-we-go-now-63957
2. See coroner's inquest reports into the Lakanal House fire, www.lambeth.gov.uk/elections-and-council/lakanal-house-coroner-inquest
3. See Greater Manchester Fire and Rescue Service: Evacuation, paras 156–65, *The Cube, Incident Report and Key Observations Regarding the Fire Which Occurred on 15 November 2019*, www.manchesterfire.gov.uk/media/2118/cube_report_v11_tagged.pdf
4. See *Simultaneous Evacuation Guidance – Guidance to Support a Temporary Change to a Simultaneous Evacuation Strategy in Purpose-built Blocks of Flats*, October 2020, www.nationalfirechiefs.org.uk/write/MediaUploads/NFCC%20Guidance%20publications/Protection/Simultaneous%20evacutation/Simultaneous_Evacuation_Guidance_october_2020.pdf
5. The Management of Houses in Multiple Occupation (England) Regulations 2006, Regulation 4(2)
6. The Smoke and Carbon Monoxide Alarm (England) Regulations 2015, Regulation 4
7. See Fire Industry, *Association Guidance for Responsible Persons on False Alarm Management of Fire Detection and Alarm Systems FIA Guidance for the Fire Protection Industry*, www.fia.uk.com/static/uploaded/385296de-a54b-443c-ac870d6651636027.pdf
8. Manchester Evening News, *Bolton Fire Students Say They 'Ignored Alarm Because It Goes Off All the Time,' as Video Emerges Showing Early Stages of Blaze*, 16 November 2019, www.manchestereveningnews.co.uk/news/greater-manchester-news/bolton-fire-students-say-ignored-17268038
9. See BS 5839–6:2019, Fire detection and fire alarm systems for Buildings Part 6 Code of Practise for the design, installation, commissioning and maintenance of fire detection and fire alarm systems in domestic premises clause 25: routine testing
10. Home Office, *Fire Statistics Table 0601: Primary Fires in Dwellings and other Buildings, by Cause of Fire, England*, www.gov.uk/government/statistical-data-sets/fire-statistics-data-tables#incidents-attended
11. *Fire Statistics Table 0602a: Primary Fires by Source of Ignition and Whether the Cause was by Human/non-human Factors 2, England*, www.gov.uk/government/statistical-data-sets/fire-statistics-data-tables#incidents-attended
12. Electrical Safety First, *Overloading Sockets*, www.electricalsafetyfirst.org.uk/guidance/safety-around-the-home/overloading-sockets/
13. British Standards Institute BS 7671:2011 – IET Wiring Regulations, 17th edition, third amendment
14. Landlords in Wales must comply with Building Regulations 2010, the Electrical Equipment (Safety) Regulations 1994 and the Renting Homes (Wales) Act 2016. Private landlords in Scotland must comply with the Housing (Scotland) Act 2006. Northern Ireland has no specific electrical safety legislation for the privately rented sector. However, properties must be fit for human habitation.
15. *The Association of Manufacturers of Domestic Appliances. Safety Repairs and Recall*, www.registermyappliance.org.uk/products/recall-list/
16. See clause 5.4.3: BS 5306–8:2012 – *Fire Extinguishing Installations and Equipment on Premises. Selection and Positioning of Portable Fire Extinguishers. Code of Practice*

6 Fire safety, the Housing Health and Safety Rating System and property licensing – law and practice

6.01

LHAs have an important role to play in ensuring that the risk of uncontrolled fire is reduced in dwellings of all sizes and tenures. In England and Wales, the HHSRS regulations designate fire as a hazard for the purpose of s.10 of the Housing Act 2004 ('the Act').[1] These are used by LHA officers to risk-assess the hazard in domestic dwellings and identify what existing and additional fire precautionary facilities may be necessary to reduce the fire hazard to as low a level as possible.

6.02

Historically, the main focus of the work of LHA officers has been to improve fire safety provisions within the private rented sector, mostly in HMOs and other rented houses. Their key reference point in seeking to make improvements has been the Act and its earlier iteration, the Housing Act 1985. Property licensing began in 2006. This is a complicated system that involves:

- mandatory licensing of larger HMOs;
- discretionary powers allowing LHAs to introduce:
    - additional HMO licensing schemes for smaller HMOs;
    - selective licensing schemes for other private rented houses.

Among other things, the purpose of property licensing is to improve fire safety provisions and management within the private rented sector. In London alone, around half of the boroughs have introduced an additional HMO licensing scheme; a third of boroughs have established selective schemes. Other boroughs regularly consult on new schemes locally.[2]

6.03

The HHSRS and property licensing are linked, but their functions are found in separate parts of the Act. An HHSRS assessment is not part of the licensing process. The management of HMOs is controlled by regulations that are also a tool used by officers to ensure correct management of fire precautions. These different pieces of law can be used together to reduce the overall fire hazard. However, officers need to have a clear understanding of how they work separately so as not to confuse the functions. This is important because the powers of entry and the means of enforcement are different.

HHSRS: hazard of fire

6.04

In England and Wales, in securing improvements to ensure adequate fire safety provisions for the many HMOs and rented houses that do not require licensing, officers must follow the HHSRS procedure. (This does not apply to other parts of the UK.) In parts of England where there is no additional HMO licensing scheme in operation, it can be difficult for officers to identify smaller HMOs because landlords are not under an obligation to notify the LHA. By contrast, in Wales all private landlords must register, and all managers must secure a licence under the Rent Smart Wales scheme.[3]

6.05

The hazard of fire is described in the HHSRS operating guidance under Profile 24 – Fire, which covers threats from exposure to uncontrolled fire and associated smoke at a dwelling.[4] The profile enumerates factors that can contribute to the likelihood of a fire. These include defects arising from electrical installation that may lead to ignition – for example, a short circuit or defects to the electric meters, sockets or switches. When seeking to take enforcement action to reduce the fire hazard in a dwelling, officers must identify the risks and use the fire hazard profile. In September 2019, the First-tier Tribunal (Property Chamber) quashed an improvement notice where deficiencies to the electrical installation found to be more relevant to electrical hazard (Hazard Profile 23) were used as a reason to serve an improvement notice to reduce the fire hazard.[5]

6.06

Officers are required to undertake a whole-house HHSRS assessment to identify all hazards, including fire. For multi-occupied properties, the assessment is made for each individual dwelling, the associated common parts and proximity to the escape route. The issue of proximity to the escape route may produce variant rating outcomes for different dwellings within the same property (for example, for different bedsits in a block).[6] Where an officer identifies an unacceptable fire hazard that needs to be reduced, they should follow their authority's enforcement policy when considering what action to take and the vulnerability of any occupiers. This may include an option for:

- informal engagement with landlords;
- the canvassing of the wishes/views of landlords and tenants;
- consultation with other housing support services (such as tenancy support and housing needs).

The HHSRS enforcement guidance states that *'authorities are likely to find formal enforcement particularly important in the case of rented properties and HMOs in the private sector.'*[7] Where a category 1 hazard has been identified, the LHA has a duty to take action to reduce the hazard. The use of informal action may leave vulnerable tenants at the mercy of retaliatory eviction and living in unsafe houses until such time as formal action is taken.[8] Officers must retain thorough and comprehensive records of evidence, including file notes of conversations with relevant parties. This is vital because, as part of the enforcement process, a Tribunal will expect full, relevant documentation to be available.

6.07

When an officer serves an improvement notice or prohibition order, they must follow the procedures contained in Schedules 1[9] and 2[10] of the Act. In serving notice, they must establish with certainty who the recipient of the notice should be, a task that is not always straightforward because of complexities around who is responsible for different parts of a building. A recipient of a notice or order has the right of appeal to the First-tier Tribunal in England and Wales.

LACORS guidance: fire safety guidance for existing housing

6.08

Originally published by LACORS in 2008, this non-statutory guidance was developed in partnership with a range of organisations with government support.[11] The guidance applies to England but may be relevant to

Wales. Similar guidance was published in Scotland in 2018.[12] The guidance adopts a risk-based approach to fire safety and takes as its primary frame of reference compliance with the Housing Act 2004 and the Regulatory Reform (Fire Safety) Order 2005. It includes guidance to landlords on the general principles of fire risk assessment and fire precautions management. The guidance is also useful in supporting LHA officers to discharge their responsibilities evaluating any residual risk where improvements are required (including the installation of fire alarms or fire doors). The document provides guidance for certain types of existing residential accommodation, including single family dwellings, shared houses, bedsits and flats. It is not aimed at housing built to meet the Building Regulations 1991 unless 'occupied in a manner other than intended under the original construction or conversion.' A clarification document was published in March 2009 to help users adopt a risk-based approach to defining risk (particularly in respect of shared houses) and the need for protected fire escape routes, escape windows and fire safety in single family houses. The document also advises users not to apply the case studies as a prescriptive standard.

6.09

Within the sector, the guidance is an important touchstone. Given its status, officers are advised to frame their case with close reference to the guidance when making submissions to Tribunals sitting to consider appeals against enforcement notices.[13] All practitioners – landlords, managers and LHA officers – continue to rely substantially on the guidance, but it is now in need of reviewing and updating to reflect the changes in fire safety policy and current British Standard. Notwithstanding the importance of the LACORS guidance, officers should resist applying its case studies to all similar types of dwellings without taking full account of the specifics of each real case. Nor should they use the case studies to require unnecessary additional fire precautions beyond existing arrangements, if these are, in reality, adequate. Officers should assess the existing precautions and require improvements or apply licence conditions where necessary to reduce the fire hazard.

6.10

The existing published fire safety guidance documents are broadly similar across the UK. In England and Wales, the LACORS guidance is the key reference document; similar guidance is available in Scotland[14] and Northern Ireland.[15] Specific regulations have been published for the Isle of Man[16] and fire certificates are required in the States of Jersey.[17] Landlords and officers must make sure that they refer to the correct national guidance. The following paragraphs in Chapter 7 provide a checklist of what are typically appropriate and adequate passive and active precautions across a range of rented property arrangements.

Consultation with the fire services on HMOs

6.11

Where, using HHSRS, LHAs identify a fire hazard in an HMO or within the common parts of a block of flats, they have a duty to consult with the local fire and rescue service under s.10 before taking any enforcement action. Where emergency action (emergency remedial or prohibition) is judged necessary, the requirement to consult applies only insofar as it is practical to do so before commencing enforcement measures. Failure to consult properly with the service can result in the quashing of a notice by the First-tier Tribunal – for example, in an appeal against an improvement notice served by Wycombe District Council, October 2019.[18] The Tribunal was satisfied that some improvement work was necessary, but, because mandatory consultation had not been undertaken, the notice was ineffective.

HMO licensing: Housing Act 2004 – Part 2

6.12

Mandatory nationwide HMO licensing schemes in England and Wales were introduced in 2006. One of the primary focuses was on improving fire safety. Independent research undertaken by Entec for the UK Government in 1997 indicated that the risk of dying in a fire was six times higher in a house of bedsits than in a family house.[19] The risk of a fire starting and leading to injury in a smaller shared house is generally considered to be broadly similar to that for a family house. However, a small shared house that does not require a mandatory licence may require an additional HMO licence in some LHA areas where discretionary schemes have been introduced.

6.13

Typically, HMOs that fall outside mandatory licensing schemes include:

- HMOs occupied by four or fewer people in more than one household;
- cluster flats in PBSA blocks managed by a private provider;
- poorly converted blocks of flats that fall into s.257 of the Act (see Para 7.26).

However, some of these HMOs may require an additional HMO or a selective licence where LHAs have introduced discretionary licensing schemes in England.

6.14

The scope of this monograph does not extend to articulating the complex meanings of HMOs in detail or to surveying which types of HMO fall into the various licensing schemes. Instead, the reader is directed to the various national regulations for clarification. However, it is important to record that there is variance within the UK on which HMOs require licences. In England, the mandatory licensing threshold was extended in 2018 to include all properties occupied by five or more people in two or more households. Meanwhile, in Wales, mandatory licensing is required for HMOs that are three storeys high and occupied by five or more people. In Scotland, all HMOs require a licence, irrespective of the number of storeys. Jersey, Guernsey and the Isle of Man have individual laws to license or control HMOs.

6.15

Regardless of whether an HMO or excepted accommodation (which falls outside the definition completely) needs a licence or not, all rented accommodation requires some elements of fire precautionary facilities, based on the risk to effectively reduce the hazard. These fire precautions must be effectively managed.

6.16

An LHA cannot grant an HMO licence until it is satisfied that:

- it is reasonably suitable for occupation or can be made so under s.64[20] to meet the standards prescribed under s.65;[21] or
- it can be made reasonably suitable through the imposition of licence conditions under s.67 (which can relate to fire safety).[22]

The prescribed standards contained in separate regulations stipulate that '*appropriate fire precautionary facilities and equipment must be provided of such type, number and location as is considered necessary.*'[23] No further guidance has been given to help LHAs understand what fire precautionary facilities are expected, but, when installing these facilities, LHA officers must give full consideration to the overall fire risk presented by the HMO.

6.17

LHAs are not legally entitled to seek to make landlords install fire precautionary facilities informally before licensing the HMO. This would, in effect, amount to making these improvements a precondition of granting a licence. The correct procedure for LHAs is to impose conditions that give the licence holder a time period to comply or alternatively to appeal if they disagree with the conditions. This prevents officers from requiring the imposition of disproportionate fire precautions before they grant the licence. However, at an early stage, sensible conversations about what fire precautions are necessary can only be helpful if they are risk-appropriate.

Licensing conditions for all property licences

6.18

English and Welsh LHAs must impose certain mandatory conditions when granting either an HMO or selective licence. The only mandatory condition relating to fire safety has to do with smoke alarms: their installation and proper maintenance plus a declaration to the local council about their positioning and location.

The only mandatory conditions relating to fire safety are that:

- smoke alarms are installed;
- they are kept in proper working order;
- a declaration is made to the local council on the positioning and location of the alarms.[24]

Licensing student accommodation

6.19

The Act excepts certain dwellings from the meaning of HMOs under Schedule 14(4). Excepted dwellings include certain buildings occupied by students.[25] Specifically, student accommodation buildings are excepted only if:

- they are managed or controlled by specified educational establishments (usually universities and colleges);[26] *and*
- the educational establishment is a member of a code of practice approved under s.233 of the Act.[27]

Officers should be clear that this is a two-part test: not all student accommodation is excepted from licensing. In excepted buildings, the

HHSRS still applies and a fire risk assessment may be necessary in the common parts under the Regulatory Reform (Fire Safety Order) 2005.

Management of Houses in Multiple Occupation regulations[28]

6.20

Management regulations have been made to place duties on the person managing an HMO in respect of repair and maintenance and equipment in the property.[29] They also impose formal duties on the occupiers.[30] The regulations apply to all HMOs, irrespective of whether a licence is required. Regulation 4 provides that escape routes must remain clear and free of obstruction and that fire-fighting equipment and fire alarms should be maintained in good working order.

6.21

Occupiers have a duty to follow reasonable instructions made by the manager to keep escape routes clear and not to tamper with fire-fighting equipment (for example, removing self-closing devices or covering alarms). Managers should issue clear fire safety instructions to occupiers and conduct routine management inspections of the common parts, including regular testing of the alarms. The regularity of inspections is a matter for the manager to determine, but they may be guided in this by LHA officers. It is good practice for managers to maintain log book records of the commissioning, testing and servicing of alarms to lay an audit trail that demonstrates good fire safety management. These records should be supplemented by photographs and messages to occupiers where there is evidence that their behaviour is compromising fire safety – for instance, where cycles parked in a hallway impede evacuation in the event of fire, or where electrical cables are trailed across a landing. At the time of writing, there are no known cases of occupiers having been prosecuted under Regulation 10. Generally, it is the manager who is pursued, especially where they are not in a position to provide a defence.

6.22

Compliance with the regulations is strict: a manager has either met the regulations or they have not. If management failures are established, the managers can expect to be prosecuted for breaches. If a manager wants to challenge a breach, their only recourse is to enter a plea of not guilty and argue their case at trial in a Magistrates' Court.

The inter-relation of HHSRS, licensing and management regulation

6.23

The various parts of the Act and regulations overlap, especially in licensed properties. This can lead to confusion. In this context, officers need to exercise considerable care in marshalling their use of the legislation in order to prevent procedural errors that may result in a Tribunal quashing a notice or order.

6.24

LHAs should seek to reduce or remove a fire hazard as soon as reasonably practicable by using the HHSRS and not by means of licence conditions, particularly where there is a significant hazard.[31] However, officers may impose special licence conditions relating to the installation or maintenance of facilities or equipment to reduce a fire hazard, even if the same result could be achieved by the exercise of Part 1 functions.[32] This approach allows for proportionate additional fire precautions to be installed.[33]

6.25

Officers cannot use HMO management regulations to secure improvements – for example, the installation of alarm systems and fire doors in HMOs – as there is no notice provision where a manager can appeal. Improvements can only be made using the HHSRS. In many instances the landlord is not the manager and may not have the responsibility or authority to install the improvements: managers can only manage what is provided.

6.26

The HHSRS and licence conditions are important regulatory tools to help achieve the primary objective of reducing fire hazards. Landlords and managers need a clear prior view of how they intend a house to be occupied: is it going to be let to a pre-formed group sharing the house for a fixed period, or is each room going to be let individually? Will the occupiers be sharing a kitchen, or will individual lessees be using exclusive cooking facilities in their rooms? Answers to the questions will affect the outcomes of fire risk assessment and any follow-on remedial actions.

Once the landlord or manager has a clear view on how the house will be let, occupied and used, they must decide on the type, standard and coverage of fire precautions that need to be installed to reduce fire hazard. Their actions should be guided by what is proportionate, and due account of a range of factors needs to be taken. These factors include:

- property size;
- existing passive and active precautions;
- protection offered by escape routes;
- the numbers of units that have individual cooking facilities;
- numbers and vulnerabilities of tenants;
- the standard of property management.

Early discussions between officers (or a suitably competent assessor) on the one hand and the landlord or manager on the other will go a long way to minimising the chance of incorrect installations that may need to be changed at a later stage. All parties need to be aware of the correct legal procedures as they engage with each other.

Notes

1. See the Housing Health and Safety Rating System (England) Regulations 2005, SI 3208 and the Housing Health and Safety Rating System (Wales) Regulations 2006, SI 1702 (W164); Regulation 4 prescribes a fire hazard if the risk of harm is associated with exposure to uncontrolled fire and associated smoke
2. See www.londonpropertylicensing.co.uk – an information website that provides details of all licensing schemes and consultation in London
3. The Housing (Wales) Act 2014
4. See Hazard 24 – Fire, *Housing Health and Safety Rating System – Operating Guidance*, ODPM Publications 2006
5. See First-tier Tribunal decision: BIR/41UD/HIN/2019/0007 & 8 – Little Meadow, St Chads Close, St Chads Road, Lichfield, WS13 7LZ
6. See LACORS Housing, *Fire Safety*, Para A.15, www.cieh.org/media/1244/guidance-on-fire-safety-provisions-for-certain-types-of-existing-housing.pdf
7. Office of the Deputy Prime Minister, *Housing Health and Safety Rating System, Enforcement Guidance*, 2006, Para 2.16
8. See further discussion about informal action in S. Battersby and J. Pointing, *Statutory Nuisance and Residential Property*, 2019, Para 5.49–5.53
9. Housing Act 2004 Schedule 1: Procedure and appeals relating to improvement notices made under s.18
10. Housing Act 2004 Schedule 2: Procedure and appeals relating to prohibition orders made under s.27
11. LACORS Housing, *Fire Safety: Guidance on Fire Safety Provisions for Certain Types of Existing Housing*, www.cieh.org/media/1244/guidance-on-fire-safety-provisions-for-certain-types-of-existing-housing.pdf

12 Scottish Government, *Fire Safety Guidance for Existing Premises with Sleeping Accommodation*, www.gov.scot/publications/practical-fire-safety-guidance-existing-premises-sleeping-accommodation/
13 See Upper Tribunal decision by A.J. Trott: 'The LACORS guidance is clearly important and ought to be given great weight in a case such as this. It appears that the guidance was not drawn to the RPT's attention, and it cannot be criticised for not referring to it. But the council undoubtedly should have drawn the report to the tribunal's attention and it is appropriate in these circumstances that permission to appeal should be granted.' UT Neutral citation number: [2012] UKUT 194 (LC), LT Case Number: HA/4/2011 – Flat 23B, Witham Road, Woodhall Spa, Lincolnshire, LN10 6RW
14 Scottish Government publication, *Practical Fire Safety Guidance for Existing Premises with Sleeping Accommodation*, June 2018
15 HMO Fire Safety Guide, *Information on Complying with Fire Safety Law in Northern Ireland*, Version 1, October 2019
16 Isle of Man, *Fire Precautions (Houses in Multiple Occupation and Flats) Regulations 2016*, Statutory Document No. 2016/0218
17 Fire certificates are issued by Jersey Fire and Rescue for designated premises including HMOs
18 CAM/11UF/HIN/2019/0006, 261 Hithercroft Road, High Wycombe
19 House of Commons Library briefing paper Number 0708, Houses in Multiple Occupation (HMOs) England and Wales
20 Housing Act 2004 S64(3)
21 Housing Act 2004 S65(1)
22 Housing Act 2004 S67
23 SI 2006, No. 373, Housing, England, the Licensing and Management of Houses in Multiple Occupation and Other Houses (Miscellaneous Provisions) (England) Regulations 2006
24 Housing Act 2004, Schedule 4(4), made under Sections 67 and 90
25 Housing Act 2004, Schedule 14(4), buildings occupied by students
26 Regulations contain a list of specified educational establishments, most recently the *Houses in Multiple Occupation (Specified Educational Establishments) (England) Regulations 2019*, No. 904, with similar Welsh regulations SI 2006, No. 1707 (W169)
27 Membership in either the *ANUK/Unipol Code of Standards for Larger Developments for Student Accommodation Managed and Controlled by Educational Establishments* or the *Universities UK/Guild HE Code of Practice for the Management of Student Housing* excepts buildings from the meaning of HMO
28 *The Management of Houses in Multiple Occupation (England) Regulations 2006* No. 372 are made under s.234 of the Housing Act 2004. There are similar regulations made in Wales, No. 1713 (W175)
29 Housing Act 2004, s.234 2(a)
30 Regulation 10 imposes duties on the occupiers of HMOs to 'comply with the reasonable instructions of the manager in respect of any means of escape from fire, the prevention of fire and the use of fire equipment.'
31 Housing Act 2004, s.55(5)(c)
32 Housing Act 2004, s.67(4): licence conditions
33 See LACORS Housing, *Fire Safety*, Paras A.40–A.43

7 Fire risk and precautions for different types of housing

7.01

Whether a rented property is occupied by a large or small single family, a group of sharers or diverse individuals occupying separate rooms, landlords have an obligation to make sure the fire risk in the accommodation is as low as possible. The landlord or manager is responsible for providing appropriate fire precautions and for advising and educating the occupiers as necessary to keep the fire risk low. Understanding precisely what fire precautions are needed and what information to give occupiers can be complex. Every house is different, as are the occupiers. For many landlords and occupiers, fire safety is limited to an awareness of smoke alarms, which may only be tested occasionally, or which may activate during cooking. Correctly installed passive and active fire precautions are integral fire safety components necessary to protect the lives of occupiers in their homes, particularly when they are asleep.

7.02

Fully functioning smoke alarms have been shown to reduce the number of fire deaths and serious injuries. To do their job effectively and reliably, smoke alarms need to be correctly installed and regularly tested and maintained. The *English Housing Survey, Fire and Fire Safety Report 2016–17* underlined the importance of working fire alarms and reinforced the case for fitting fire doors to kitchens in privately rented houses:[1]

> *In 2016–17, 332,000 households in England had experienced a fire at home in the last two years. In the majority of these households, the fire was put out by someone in the household, or the fire went out by itself.*
>
> *Since 2015–16, the proportion of private renters with a working smoke alarm has increased from 83% to 88%. Of those fires that started inside the*

house or flat, two thirds (67%) started in the kitchen. It is therefore not surprising that fires in the home were most commonly caused by cooking related activities, such as a grill or chip pan catching fire.

7.03

In England, for the year ending September 2019, of the 28,655 primary dwelling fires attended by fire and rescue services, just under three quarters (73 per cent) were in houses, bungalows, converted flats and other properties. The rest occurred in purpose-built blocks of flats. This suggests that houses are not necessarily safer than purpose-built blocks of flats.[2] Many HMOs are not purpose-built – for example, older buildings converted into self-contained flats and newer houses originally intended to be used as single family houses. Some of these will not have had adequate fire detection or compartmentation installed at the time they were converted (particularly older properties). For this reason, they will not take proper account of the change of risk arising from different occupation. When renting out a property, landlords may have to upgrade the original fire precautions to a higher standard to reduce the fire risk or to meet property licensing conditions.

7.04

Chapter 3 introduced how a fire can start and spread within dwellings, and Chapter 4 discussed the importance of adequate and correct passive and active precautions. All dwellings are different, and factors that affect fire risk include:

- building height;
- layout;
- style of occupation;
- numbers of kitchens.

The purpose of this chapter is to help all those associated with rented housing (including landlords, managers, tenants and regulators alike) to consider how best to reduce fire risk through the installation of additional passive and active precautions (such as compartmentation, alarms and means of escape). Observations made here should not be taken as an alternative to undertaking a thorough fire risk assessment or engaging a competent person to do so, if desired. This chapter's scope does not include the communal areas or management of larger blocks of flats (PBSA). These will be considered in Chapter 8.

Alarm grade and category of system

7.05

A suitable and sufficient risk assessment will indicate what grade of alarm and category of coverage will be necessary to reduce the risk of fire. Table 4.1 of the BS 5839–6:2019 code of practice sets out recommendations for the minimum fire alarm systems to be installed for life protection in a typical range of different classes of premises with reference to differing floor areas and numbers of storeys.[3] Accompanying guidance notes in the code of practice may help a landlord decide the most appropriate grade and category of alarm system for a particular type of use. It is worth noting that the recommended grade and category of alarms contained in the British Standard differ in part from the existing LACORS guidance (for example, in their approaches to lower-risk shared houses). This may result in LHAs changing their guidance documents to reflect the British Standard rather than the LACORS guidance when specifying fire precautions. In the event of any enforcement action brought by LHAs, only the courts and Tribunals can adjudicate on any differences between the documents as raised in the course of a hearing or appeal.

Grade of system

7.06

The grade of fire alarm indicates the standard of detection, control and monitoring provided. Grades range from a single-point, battery-operated smoke alarm (Grade F) to an integrated, panel-controlled system (Grade A). Each grade is designated with a letter A, C, D (1 and 2) and F (1 and 2).[4] Grades B and E are currently not defined. The most common grades likely to be used by residential landlords are:

- Grade A – a panel-controlled fire detection and alarm system with control and indicating equipment that generally includes break-glass call points;
- Grade D1 – a system of one or more mains-powered detectors fitted with a tamper-proof battery standby power supply;
- Grade D2 – a system of one or more mains-powered detectors fitted with a user-replaceable, tamper-proof battery standby power supply.

Other grades may be acceptable when supported by a thorough risk assessment.

Category of system

7.07

The category of system to be installed to protect life in domestic premises has a designation of LD. Within the LD category, there are three classifications differentiated by the level of protection they provide:[5]

- LD1: detection provided throughout the premises, including all parts of the escape route areas, all rooms and areas except those with negligible sources of ignition (for example, toilets and bathrooms);
- LD2: detection provided throughout the premises, including all parts of the escape route and all specified rooms or areas that present a high fire risk to the occupiers (including kitchens and the principal habitable room or rooms);
- LD3: detection in all circulation areas that form the escape route. This is the category of alarm that is provided in all new-build houses.

Sound levels

7.08

Fire alarms must be loud enough to rouse the principal occupiers from sleep and give sufficient early warning to allow escape. A sound level of 75dB(A) at the bedhead is generally accepted to be sufficiently loud to rouse most sleepers. This is similar to the sound level generated by many vacuum cleaners. However, there is no guarantee that people heavily under the influence of alcohol or drugs will be woken. Closed doors – particularly fire doors and other solid doors – can reduce sound levels. Given that many people sleep with their bedroom door closed – particularly in shared housing – it is often necessary to install supplementary alarms in bedrooms, linked to the existing alarms. Although children may sleep through an alarm, it is generally assumed that parents/carers would wake their children in the event of an alarm sounding. Following the deaths of six children in a deliberate house fire in Derby in 2012, research by Dundee University and Derbyshire Fire and Rescue found that, of 34 children tested, 27 repeatedly slept through smoke detector alarms.[6]

Separation, partitions and escape routes

7.09

The purpose of adequate compartmentation has already been discussed. Uncontrolled smoke and hot gases can spread rapidly through fire separation by seepage through the tiniest of gaps. When letting or altering a house, landlords are directly responsible for ensuring adequate separation to protect all the occupiers, guests and those in neighbouring dwellings. Where dwellings have been converted from their original to a new use, the installation of additional fire separation works may be necessary. For example, transforming a family house into HMO accommodation by creating extra bedrooms may necessitate hanging fire doors and erecting new partitions. It is essential that these are properly sealed and a good fit. Where walls are pierced to accommodate new cabling, special attention must be given to properly sealing the holes in order to stop smoke and hot gases percolating between living units.

7.10

It is possible that some refurbishments or conversions do not need planning permission or building control consent. In these cases, in the absence of third-party expertise, landlords may not understand what additional fire separation may be required to reduce the fire risk. This may result in inadequate fire separation, for example, not hanging fire doors where necessary, poor workmanship resulting in large holes left in compartmentation or the use of incorrect materials, such as the wrong type of mastic or other filler (see Image 7.1).

Image 7.1 Expanding foam used to fill excessive gaps in a separating wall. Para 7.10

Number and location of fire doors in rented houses

7.11

Additional fire doors may not always be needed, especially in two-storey, single-occupancy houses and lower-risk shared houses. They will, however, always be necessary in bedsit accommodation and converted flats. The precise number and location of fire doors needed will depend on the internal layout and room use: the primary purpose of a fire door is to protect the escape route from becoming filled with smoke and hot gases. The importance of correct installation of fire doors by competent fitters cannot be overstated. Fire doors that are a bad fit will not work as designed. They must be thoroughly and regularly inspected and maintained. Landlords need to be vigilant when engaging and overseeing contractors to do fire compartmentation work. Having to rectify substandard work at a later stage may prove expensive. It is possible that poor or incorrect separation may only be identified as a result of an LHA or fire service inspection, triggered as part of a licensing regime or in response to a complaint.

7.12

Lightweight, thin, composite internal doors, commonly referred to as 'egg box' doors, often found in new-build houses, will offer very little fire protection and can be damaged easily. Where these types of doors are provided as kitchen doors, consideration should be given to replacing them with a more substantial, solid core door.

Wall and ceiling linings

7.13

Wall and ceiling linings should not support the rapid spread of flames. For example, the use of polystyrene ceiling tiles or thin wooden panelling should be avoided (unless the wood has been treated with a fire-retardant lacquer) – particularly in fire escape routes (see Image 7.2).

7.14

When polystyrene ceiling tiles melt, they release toxic gases and drop molten plastic, which can cause further ignition and burns. Currently, they are not illegal in England, but they are certainly hazardous. This

Image 7.2 Polystyrene ceiling tiles. Para 7.13

anomaly prompted Graham Morris MP to put a parliamentary question on 21 June 2017:

> *To ask the Secretary of State for Communities and Local Government, if he will bring forward legislative proposals to prohibit the use of polystyrene ceiling tiles in rented accommodation.*[7]

Alok Sharma MP, Minister for Housing, replied that local authorities were able to use HHSRS enforcement powers to require their removal. Clearly, in order to do this, the LHA has to be made aware of the tiles, identify them as a fire hazard and take appropriate enforcement action. It may not be easy to identify the existence of ceiling tiles until, for example, a house requires a licence, or a tenant makes a complaint.

Escape route lighting

7.15

Depending on the type of tenancy, it is likely that emergency lighting will be needed where an escape route is long, complicated or has no borrowed lighting. A three-storey terraced townhouse with an escape route that has no windows and a winding staircase is likely to require some element of emergency lighting if occupied as an HMO. This is less likely if the property is occupied as a family house. By contrast, a two-storey house with a straight staircase and windows at the first-storey landing will probably not require emergency lighting, even when occupied as an HMO.

7.16

Landlords are responsible for ensuring there is effective conventional lighting in the escape routes in the common parts of HMOs. Where timer switches are provided, illumination should last long enough to allow occupiers to escape to a place of safety. On occasion, tenants may take the lightbulb from the escape route as a replacement for a blown bulb in their accommodation. The provision of different bulb holders in the bedrooms and the escape route will remove this temptation; for example, a bulkhead light or screw-fitting to the pendants in the escape route and bayonet fittings in the bedrooms.

Open-plan layouts and inner rooms

7.17

In university areas, there are many examples of family housing originally built with full-depth living rooms subsequently converted for use as shared student accommodation. The living rooms are divided centrally to provide an extra ground-floor bedroom. This creates an inner room with access from an outer room, for example, the kitchen. In the event of a fire in the outer room, the occupier of the inner room must have a secondary means of escape and sufficient time to get out before being overcome by smoke or hot gases. To be acceptable, the separating door between the outer and inner rooms needs to be a 30-minute fire door, and a window must be installed that meets the standard of an emergency escape window to Building Regulations requirements.[8] This will allow secondary escape to a place of safety in the event of a fire in the outer room. A secondary escape window is only acceptable up to a height of 4.5m from the ground (that is, no higher than first-storey level) and if it is easily accessible and available for use by all occupiers or guests who may need to use it. In 2018, the First-tier Tribunal considered the facts of an escape window located in a loft room with a sloping ceiling in a bungalow. To use the escape window, an occupier would have had to slide down 2m of a tiled pitched roof onto a flat roof before dropping to the ground. In dismissing an appeal against a prohibition order, the Tribunal observed that:

> *If exit were required in the circumstances of an emergency, it would most likely entail throwing oneself headfirst out of the window onto the pitched roof and sliding approximately 2m headfirst until reaching the flat roof.*[9]

7.18

Emergency escape windows must meet seven criteria to achieve the standard set out in the LACORS guidance. Officers need to exercise professional judgement to reach a determination on the suitability or otherwise of an escape window against each criterion. As part of a successful appeal to the Upper Tribunal in 2012, careful and detailed consideration was given to each of the criteria. The appeal was against the decision of the Residential Property Tribunal to confirm a prohibition order served on a first-floor, one-bedroom flat on the grounds that the layout was unsatisfactory despite the escape windows complying with the Building Regulations 2010. The Building Regulations now permit removable keys and stays to emergency escape windows. This differs significantly from the LACORS guidance. In his judgement, A.J. Trott FRICS concluded that '*in my opinion the Building Regulations should take precedence over the LACORS guidance.*'[10]

7.19

Open-plan living arrangements are increasingly popular in the UK. One of their perceived advantages is that they can maximise available floor space. However, they clearly pose an additional risk to safe escape. Many open-plan dwellings do not have a protected entrance hall or lobby as the stairs end in the living room, which forms the fire escape route. Habitable rooms located off the living room (usually bedrooms) are then considered an inner room. Generally, open-plan layouts with inner rooms are considered appropriate only for lower-risk, single family occupation, shared houses and flats: there is generally more communication and interaction between occupiers, who are therefore more likely to know who is in and who is out. This increases the chances of safe escape. The current LACORS guidance does not directly address open-plan configuration. LHA officers encountering open-plan arrangements may well decide that these layouts present too great a hazard. They may seek either to prohibit the use of all, or part of, the dwelling or to require the installation of additional protected escape partitions, which may in reality be impractical or impossible.

7.20

There are, however, some alternative solutions to these difficulties. National House Building Council (NHBC) research has concluded that for a specific and limited size and design of open-plan flat, the installation of sprinklers combined with an enhanced alarm system can provide an acceptable level of safety.[11] As always, the ongoing testing of the active fire precautions are critical to demonstrating that the management is good enough to reduce the risk.

Local guidance documents

7.21

Most LHAs have produced local fire guidance documents, many of which are based on national guidance (for instance, the LACORS guidance to help landlords and managers understand what type of precautions may be necessary). In order to promote local consistency, some LHAs collaborate with neighbouring authorities to convene county-wide officer groups that produce shared documents. This is often helpful for managers with large property portfolios that straddle authority boundaries. It is important that these guidance documents are reviewed regularly to reflect shifting local conditions and to address any recent Tribunal decisions that may be relevant.

The following paragraphs work through examples of what fire precautions may be necessary in different types of buildings with references to recent case decisions made by either the Residential Property Tribunal (later the First-tier Tribunal) and the Upper Tribunal that responsible persons and officers may find of use.

HMOs

7.22

HMOs include a multitude of buildings with different layouts, construction ages, storeys and numbers of kitchens. All sizes and styles of HMO require a fire alarm based on the risks of ignition, fire spread, occupier numbers and behaviour. The fire risk in a house can be dependent according to occupation rather than tenancy type (see Para 4.16). Indeed, there is no legal definition of 'shared houses' or 'bedsit-type HMOs.' The LACORS guidance gives a description of each and outlines the differences in perceived fire risk presented by these very different types of HMO. Not all shared houses are low risk, and not all bedsits are necessarily high risk; so what about properties that fall between these broad types? In undertaking assessments, officers should follow guidance documents and apply what they advise to particular cases and particular risks. An assessment of the actual risk must be made by officers seeking to make improvements and must allow for deviation from a guidance document where suitable alternatives can be demonstrated to reduce the risk.

7.23

An example of acceptable HMO fire precautions that deviated from the national guidance was highlighted in an appeal against a licence condition imposed by Weymouth and Portsmouth Borough Council (WPBC) in 2019.[12] The condition stipulated the replacement of the existing interlinked alarms with a panel-controlled alarm system. The appeal brought into focus the problems that some LHA and fire officers have in assessing the actual fire risk posed. The landlord was renovating a three-storey HMO with a single shared kitchen. His plan was to let it to five individuals. The landlord approached WPBC for fire safety advice to make sure he had interpreted the LACORS guidance correctly. He fitted a mains-powered system of interlinked smoke alarms in all bedrooms and the escape route, and he installed a heat alarm in the kitchen (Grade D interlinked alarms) with a series of test buttons fitted in the hallway. This installation was of a higher specification than advised in the LACORS guidance for a shared house, but it did not meet the guidance for a bedsit HMO. At the time of the renovations, an inspecting officer indicated that he did not consider it to be unsatisfactory for the risk. As part of the licensing application, the house was visited by a fire officer who advised that a much higher standard of alarm should be installed. The reason given was that no fire risk assessment had been seen to justify the installation of the interlinked alarm. The First-tier Tribunal considered the evidence and arguments. WPBC registered their concern about the testing and record-keeping regime for the alarms. It was accepted by both sides that the HMO did not fully align with the LACORS guidance for either a shared house or a bedsit HMO. The Tribunal took the view that the Grade A alarm system was too high a specification for the risk posed. They agreed that the existing alarm system could remain *in situ* but ruled that it would require rigorous testing, supported by detailed records submitted to WPBC quarterly.

7.24

This decision brought further into question the received wisdom that bedsits are intrinsically a high fire risk and shared houses a low fire risk, something that was addressed in a LACORS clarification document published in March 2009.[13] However, the notes to Table 4.1 of BS 5839-6 make specific distinctions between houses used as HMOs and shared houses (interpreted as houses shared by no more than six people, generally living in a similar manner to a single family – for example, houses rented by a number of students).

7.25

HMOs and shared houses all require an element of passive fire precautions. Other than in a very few cases, kitchens need a fire door. Licence conditions requiring 30-minute fire door sets to be fitted to kitchens are likely to be confirmed by a Tribunal. In a case in January 2020, the First-tier Tribunal dismissed an appeal against a licence condition imposed by Bath and North East Somerset Council confirming the requirement for a fire door to be fitted to the kitchen: *'the Tribunal is satisfied that the fire door specified in the licence in this case is still of real value in protecting [the residents].'*[14]

Fire doors will also be required for any inner rooms that have access from a kitchen or living room. They may also be necessary on bedrooms, depending on the existing standard of door, other compartmentation, number of storeys and layout of the house.

Poorly converted blocks of flats

7.26

Before the Building Regulations 1991 came into effect, older houses that were large enough to convert into self-contained flats did not require rigorous standards of fire resistance between each flat (for example, adequate floor/ceiling separation or the installation of fire doors). Where one such house is still not compliant with the 1991 regulations and where more than two thirds of the flats within the property are not owner-occupied, the house – not the flats within it – falls within the meaning of an HMO under s.257 of the Housing Act 2004 in England and Wales.[15] As there are two tests to meet, a converted block can meet the definition of an HMO at certain times and not at others, for example, if a previously rented flat is sold to an owner-occupier, increasing the percentage of owner-occupied flats beyond the two thirds threshold. Whether or not an LHA includes these types of buildings in an additional HMO licensing scheme, landlords have a duty to undertake an FRA for the common parts of the building and to manage the common parts to meet the national management regulations.[16]

7.27

Where conversions were undertaken many years ago, it may be difficult to prove the standard of fire resistance between the flats unless LHAs have records of conversion details to show otherwise. A stage 1 non-intrusive risk assessment is unlikely to indicate the requisite resistance of ceiling/floor separation, and it may be necessary to install additional fire precautions, for example, a communal fire alarm system.

7.28

The provision of a common parts smoke detection system where the grade installed increases with height is likely to be a minimum improvement. A well-designed alarm system should include coverage of both the communal escape route and the individual flats to provide early warning to occupiers. A mixed system linking smoke alarms in the common parts to a single heat alarm in the separate flats will provide satisfactory early warning, while minimising accidental sounding. For instance:

- in a two-storey block, a mixed system of interlinked alarms may be satisfactory. However, it is likely that, for taller buildings, a panel-controlled integrated system will be necessary;
- fire escape routes must meet construction standards for 30-minute fire resistance and include 30-minute self-closing fire doors to each flat entrance;
- fire risk assessors should carefully consider the existing standards of separation, particularly where the flats have been vertically separated between the original staircase, so that an upper flat is accessed up the staircase and a lower flat through a separate front door in the original hallway;
- for each flat, as a minimum precaution, a fire door will be necessary to the main front doors and possibly the kitchen, depending on layout and occupancy numbers (particularly if a bedroom is located off a high-risk room such as the kitchen).

Back-to-back houses

7.29

Houses of this type were built in the industrial towns of the midlands and north of England. The first back-to-backs were constructed in Leeds in 1737; 19,500 of them are still standing and are in use.[17] The houses back directly onto each other and share a rear wall. They are built in terraces. This configuration means that they have only a single staircase fire escape route to the front of the house. In most back-to-backs, the escape route passes through a risk room, so that all the first-floor bedrooms are treated as inner rooms. In most back-to-back houses:

- depending on internal layout, a fire door will need to be installed across the staircase to separate the ground and first floor with fire escape windows provided at first-floor level;
- interlinked alarms are likely to be necessary in all habitable rooms and the kitchen to ensure early warning and good audibility levels.

Fire safety – a joint responsibility in the home

7.30

Effective fire safety in the home is a joint responsibility, involving good communication of responsibilities between both the landlords and occupiers across all sectors and tenures. Property owners are responsible for providing a safe home that reduces the risk of fire as far as possible through provision of passive and active precautions. Occupiers must make every effort not to create a fire hazard and have a responsibility to follow reasonable instructions. The following paragraphs set out a basic checklist of responsibilities for landlords and tenants. The points have not been placed in any particular order of priority. Each one makes a valuable contribution to minimising the fire hazard.

The landlord's responsibilities

7.31

Before they let a dwelling, a landlord or manager should:

- install sufficient smoke and heat alarms in the house (quantity and location to be based on risk) and test them;

- where fire doors are necessary, check to make sure they latch into the frame unaided, and replace any missing or damaged intumescent strips and cold smoke seals;
- register all new white goods with the manufacturer, and verify that the existing appliances – for example, fridge, freezer and washer-dryer – are not part of a fire safety recall campaign. Current recalls can be found on local fire service websites such as London Fire Brigade and national sites such as Electrical Safety First;
- as a legal requirement, give the tenants a current electrical safety certificate dated within the last five years;
- as a legal requirement, give tenants a current annual gas safety certificate dated within the previous 12 months;
- make sure the prospective tenants understand how to reduce the risk of accidental fires;
- make sure occupiers can unlock the doors and get out of the house quickly. This is particularly important in the case of inner rooms. Ideally, lockable doors should be fitted with a keyless opening mechanism;
- consider whether they want to ban smoking and the use of candles as a condition in the tenancy agreement and/or provide safety advice;
- check the make and model of all portable appliances in the house, provided as part of the tenancy. A portable appliance test (PAT) will only verify whether an appliance is electrically safe to use; it may still present a fire hazard;
- advise occupiers that they should only use electrical appliances that are CE-marked, and properly adapt them for correct use on the electrical supply;
- instruct the occupiers how to use any fire-fighting appliances, where provided;
- undertake regular property inspections as necessary;
- check the labels on the furniture and furnishings they supply to make sure they are fire-safe.

The tenant's responsibilities

7.32

Tenants should:

- read and make sure they understand all the reasonable instructions provided by the manager, often in the form of a welcome pack or something similar;

- make an escape plan, and be clear about the most direct escape route, especially if they have dependants and/or a secondary escape window;
- not introduce flammable items into the escape routes, and keep these as clear as possible;
- close all the internal doors at night and when the house is empty during the day – especially important for fire doors;
- clean the oven, hob, grill and grease filters regularly. Tenants should not allow grease to build up;
- not overload electrics and extension socket boards;
- avoid using candles, incense sticks, other naked flames and chip pans, whether banned in the tenancy agreement or not;
- not store white goods in the fire escape route, because they pose a fire risk that could impede escape;
- not use portable fans or convection heaters. If they do, they should not place them too close to flammable items (see Image 7.3). Small fan heaters can overheat if they are left on for extended periods.

Image 7.3 Scorched fan heater due to overheating caused by excessive use. Para 7.32

7.33

The landlord and tenants should agree on arrangements for sounding the smoke alarms once a month. The manager will remain the responsible person for making sure the alarms are tested regularly, even if the tenants agree to undertake the test by agreement.

Notes

1 Each year the English Housing Survey includes a number of questions relating to fire safety. More detailed questions on the incidence of fires at home were included in 2015–16 and 2016–17. Fieldwork concluded in March 2017, before the Grenfell Tower tragedy.

2 See *Home Office Statistical Bulletin 6/20 – Fire and Rescue Incident Statistics, England, Year Ending September 2019*

3 See Table 1, BS 5839–6:2019: Fire detection and fire alarm systems for buildings, Part 6: Code of practice for the design, installation commissioning and maintenance of fire detection and fire alarm systems in domestic premises

4 See clause 7 – BS 5839–6:2019

5 See clause 8 – BS 5839–6:2019

6 See Fergus Walsh, BBC News Report, *Smoke Alarms Fail to Wake Children*, 27 February 2017, www.bbc.co.uk/news/health-38918056

7 The full question and answer can be found at https://questions-statements.parliament.uk/written-questions/detail/2017-06-21/370

8 HM Government, the Building Regulations 2010, Approved Document B, Volume 1: Dwellings. Para 2.10: Emergency escape windows and external doors

9 Tribunal decision MAN/00EH/HPO/2018/0006, 234 Yarm Road, Darlington, County Durham DL1 1XW

10 UT Neutral citation number: [2012] UKUT 194 (LC), LT Case Number: HA/4/2011 was a successful appeal against an RPT decision to confirm a prohibition notice

11 NHBC Foundation, *Open-plan Flat Layouts – Assessing Life Safety in the Event of Fire*, September 2009

12 See Tribunal decision CHI/9UJ/HMV/2019/0005, 62 Rodwell Road, Weymouth, Dorset DT4 8QU

13 LACORS published a clarification document to assist in adopting a risk-based approach to shared housing

14 See Tribunal decision CHI/ooHA/HML/2019/0016, 12 Junction Road, Bath, BA2 3NH

15 Owner-occupiers are those with a lease of more than 21 years or who own the freehold in the converted block of flats, or a member of the household of the person who is the owner

16 The Licensing and Management of Houses in Multiple Occupation (Additional Provisions) (England) Regulations 2007 (SI 2007/1903) and the Licensing and Management of Houses in Multiple Occupation (Additional Provisions) (Wales) Regulations 2007 (SI 2007/3229 (W281)

17 See Leeds Libraries, *Back-to-Back Houses and their Communities*, May 2017, https://secretlibraryleeds.net/2017/05/22/back-to-back-houses-and-their-communities/

8 Management and regulation of blocks of flats and larger student accommodation developments

8.01

Blocks of flats and high-rise buildings everywhere are occupied by a mix of tenures, including:

- private and social tenants;
- long leaseholders who contribute a fee towards the management and upkeep of the communal parts;
- large numbers of students who are often housed in PBSA and who may be living away from home for the first time.

Different fire risks attach to different tenures and to differently constructed and configured developments. Blocks vary in their standard of construction, provision of alarms and/or sprinklers and their internal layout. Some are purpose-built; others are converted. Another critical point of specificity will be whether a block has been subject to any post-construction alterations made to its original fire separation and/or external cladding – and, if so, what type and level of alterations have been made. Building owners and managers must continually assess and reduce or manage the fire risk to protect life and property. This includes effective communication with occupiers. Regulation of fire safety in blocks of flats and larger student developments can be confusing for owners and managers. This confusion stems from the joint and mixed regulatory approach, including licensing regimes available to LHAs across England.

8.02

There is a widespread belief that living in a high-rise block of flats is less safe than living in a lower-rise block. This is not necessarily the case: high rise is not always high risk. In England for the year 2019–20, of the

28,447 primary dwelling fires affecting purpose-built blocks, low-rise blocks accounted for 4,840 fires. By contrast, there were 2,677 fires in blocks of four or more storeys.[1] Properly designed and built blocks should all have protected means of internal fire escape and adequate fire separation both to prevent fire spread throughout the building and to allow a safe and orderly evacuation. However, not all blocks have fire alarm systems or sprinkler systems – the older the property, the more likely that it does not enjoy these provisions.

8.03

Demand for more affordable housing in the UK and the need to replace houses destroyed in the war resulted in many tower blocks being built in the 1950s and 1960s, particularly by LHAs. The first residential tower block in England was built in 1951: the Lawn, Harlow, Essex, a ten-storey, reinforced concrete structure faced with brick and fitted with metal windows.[2] Uninsulated concrete-faced buildings with metal windows have the advantage that they do not support the external spread of flames; but they are not attractive, water can leak through ill fitting concrete panels as happened creating cold and damp fats that were hard to hear. This often leads to internal condensation and mould growth that adversely affects the health of occupiers. During the 1980s, the UK government imposed big increases in energy prices, especially for electricity, which is used to heat flats in tower blocks.[3] An effective response to these challenges was to insulate buildings with external cladding and a rain screen as a way to reduce heat loss and stop leaks.

8.04

It is a matter of record that, in a number of countries, there have been fires where the external cladding has been identified as a major factor in the rapid external spread of flames. In some cases, it turned out that the cladding materials did not meet the national buildings standards in place at the time. However, the Grenfell Tower Inquiry heard that on 7 July 2016 the local authority issued a building control completion certificate. In the wake of these fires, there has been a concerted effort to make the removal of external flammable cladding a priority in countries that include the UK, Australia and the United Arab Emirates. However, until the cladding and associated fire risk are removed, the buildings need to be managed and risk-assessed to protect the safety of the occupiers. Interim measures to achieve an adequate level of protection may include temporary waking watches. Alternatively, regulators may need to take enforcement action to reduce the risk.

Risk assessment: joint approach to the regulation of blocks of flats

8.05

As discussed in Chapter 4, in England, the regulation of fire safety in the common parts of blocks of flats and large student blocks is a joint responsibility born by LHAs and fire and rescue services.[4] In discharging these responsibilities, the formal frame of reference for LHAs is the Housing Act 2004, Part 1 HHSRS; and for fire and rescue services, it is the FSO. It is essential that a clear working protocol is agreed to ensure workable and consistent regulatory arrangements when applying the two sets of legislation. This should ensure consistency and prevent confusion for those responsible.

8.06

There is an absolute legal requirement for the responsible person (the landlord or manager) to undertake a fire risk assessment for the common parts of a block of flats to meet the FSO. Individual flats are exempted from the order but can be assessed by LHA Officers, using the HHSRS. The value of a risk assessment will be seriously diminished if the actions it sets out are poorly articulated to the extent that the responsible person(s) is left in any way unclear about what it is they are being asked to do and by when. Cost is not a valid reason to avoid necessary risk-reducing improvements and repairs. It is not in anyone's best interest for a competent risk assessor to identify actions that are either ignored or attempted by a person who does not understand whether or not what they are doing will reduce the risk.

8.07

Identifying the responsible person in a large building, particularly where there are multiple leaseholders, may not be straightforward as it is possible that both the owners and the managers have joint responsibilities. If the terms of the lease are unclear, it may not include the building's external structure or the individual flat front doors. When it is unclear who the responsible person is, either the owner or manager – or both – may be responsible and accountable. Either one or both may be at risk of prosecution if serious fire safety failings are discovered, as may be the developer in some circumstances. Their exposure to prosecution is especially high where residents are put at unacceptable risk, perhaps by having been allowed to occupy a partly finished building. A case in point is the court hearing held in January 2020 to consider alleged omissions by the owner, manager and developer of student accommodation in Leeds in 2016. Following admission by all parties that they were culpable for a range of failings, the West Yorkshire Fire Service observed in a statement that:

> *This case demonstrates the importance those responsible for building construction, development and occupation have in understanding their duties and acting responsibly to take account of the safety of the people they are responsible for. As Judge Mairs highlighted, the dangers and risks found at Trinity Halls were so obvious anyone without a technical fire safety background could identify them.*[5]

8.08

LHA officers who use the HHSRS when assessing the risk in the common parts of a block of flats generally inspect the individual dwelling first and then consider any relevant deficiencies in the common parts as part of the overall flat assessment. This approach leaves no doubt that the flat front door can be included in an HHSRS assessment. The Housing Act 2004 allows officers to consider the common parts. These include, for these purposes, the structure and exterior of the building and the internal and external shared facilities as separate residential premises.[6]

8.09

A suitable and sufficient fire risk assessment should include:

- assessment of the internal visible areas of a block of flats, including a sample of the flat front doors;
- consideration of the external wall covering and any construction of balconies that may contribute to the external spread of flames;
- evaluation of whether fire barriers have been properly installed in any concealed spaces such as wall cavities or ceiling voids to prevent the upward spread of smoke or flames.

This level of assessment is likely to be highly specialist in nature and requires intrusive investigation by a competent person with adequate building knowledge.

Fire doors

8.10

As discussed in Chapter 3, self-closing fire doors are a vital part of a fire compartmentation system for a tall building. These must be fitted to all front doors to flats and also be installed across longer internal escape corridors and at the top of stairs. To be effective, a fire door must fully self-close unaided into the stops without excessive gaps between the doors and the frame. Fire doors to flats are necessary to prevent a fire inside a flat breaking out into the escape route and to protect tenants sheltering within a flat from fire in the building's communal parts. This is especially important where a stay-put policy is in place. The current standard for flat doors is 30 minutes' fire protection. Failure to meet this standard is often attributable to missing combined fire strips, defective self-closing devices or replacement by leaseholders with doors that are not fire-rated. Leaseholders who want to change their flat entrance door will need to obtain both building control permission and the landlord's consent in order to make sure that the new door meets current fire safety standards.

8.11

A suitable and sufficient fire risk assessment should always include an assessment of at least a sample of the flat doors, including for leasehold flats. Assessors should be alert to any replacement doors that are not fire-resistant, and to doors whose integrity may have been compromised in some way, for example, by having been cut to accommodate a post box and flaps that are not fire-resistant. Fire-rated post-slot plates can be installed with intumescent liners to retain the standard of 30-minute protection. Where non-fire-resistant doors are identified, building owners may have to rely on the terms of the lease to secure the installation of a satisfactory door and to reclaim the costs from the leaseholders. The Upper Tribunal has held that landlords must not simply change doors that are in disrepair: there must be expert evidence that the door is unable to provide the correct level of fire resistance.[7]

External wall cladding and external balconies

8.12

External cladding attached to tall buildings has contributed to the rapid spread of flames in devastating fires across the world. The cladding in these cases has often been retrofitted to increase energy efficiency, improve visual appeal and stop leaks. Buildings that have fallen victim in this way include:

- Grenfell Tower and the Lacrosse Tower, Melbourne, both clad in the same aluminium composite material (ACM);
- Mermoz Tower, a social housing block in northern France, which was covered in flammable cladding in 2003 and which suffered a fatal fire that started on a first-floor balcony in 2012.[8]

The ongoing Grenfell Tower Inquiry found that the polythene core panels within the external ACM cladding was the primary cause of the fire spread and did not meet building control standards.[9] The independent expert advisory panel appointed by the MHCLG expressed its collective view that no wall system containing an ACM category 3 cladding panel – even when combined with limited combustibility insulation material – would have met Building Regulations.

8.13

The Cube, Bolton, a student accommodation block that suffered an externally spread fire in November 2019, had been covered in a high-pressure laminate (HPL).[10] The fire resistance performance of HPL cladding can vary widely. For this reason, it is important that building owners know what type of material has been installed. HPL is generally made of wood or paper layered with resins and bonded under heat and pressure. The fire resistance classification depends on the material's thickness and whether it has been impregnated with fire-retardant chemicals.[11]

8.14

The presence of external wall cladding is clearly a serious concern to residents occupying tall residential buildings because of the potential for the rapid spread of flames. It is essential that building owners and managers undertake a risk assessment to establish the type of cladding attached to a building. Where flammable cladding has been identified as part of a risk assessment, it must be removed without delay. LHAs can use the HHSRS to enforce the removal of flammable cladding, as well as require improvements to poor compartmentation and defective fire doors. In November 2018, MHCLG published operating guidance supplementary to existing documentation supporting the use of HHSRS. Specifically, this set out guidance for the assessment of cladding in high-rise buildings.[12] Following the Grenfell Tower fire, an interdisciplinary national Joint Inspection

Team was formed. It comprises EHOs, a fire engineer, a building control expert and a legal adviser, charged with helping LHAs with the enforcement of cladding removal where owners are reluctant to remediate.[13]

8.15

External balconies, at whatever height on a building, should be included as part of an assessment of potential fire spread. This is especially important if the construction of the balconies incorporates flammable materials. A major fire at Samuel Garside House, Barking, in June 2019 was able to spread as a result of the flammability of the wooden balconies that had previously been identified as a 'significant hazard.'[14] Residents of flats with balconies should be given the following general fire safety advice:

- not to use a barbeque or burn candles on the balcony;
- not to store flammable items such as gas bottles;
- to extinguish all cigarettes properly, and not to throw them carelessly over the balcony.

8.16

In July 2019, the UK Government launched a private sector remediation fund aimed at covering the cost of removing and replacing unsafe ACM cladding from private residential buildings higher than 18m.[15] Additional funding has been made available to fund the removal and replacement of unsafe non-ACM cladding systems, including HPL.[16] However, only buildings that are more than 18m high are eligible. At the time of writing, three years after the Grenfell Tower tragedy, there are still tall buildings in England with ACM cladding in both the private and social sectors.[17]

8.17

The amended Building Regulations 2010 came into force on 21 December 2018.[18] These prohibit the use of combustible materials anywhere in the external walls of high-rise buildings that are over 18m above ground level and that contain one or more dwellings. This ban applies to new buildings or to refurbishment work where the external wall is involved. The regulations are, however, not applied retroactively to buildings with existing cladding that needs to be assessed and remediated or removed.

External Wall Fire Review process (EWFR)

8.18

Tall building fires have demonstrated that unsafe external cladding, combustible balconies and other combustible attachments can lead to the rapid spread of fire up a building, resulting in a major disaster. The external cladding, insulation and fire breaks should work as an integrated external wall system to prevent fire spread.

8.19

In December 2019, two years after the Grenfell Tower fire, the Royal Institution of Chartered Surveyors (RICS), the Building Societies Association (BSA) and UK Finance put together a new industry-wide valuation process designed to give lenders and leaseholders confidence in the safety of external wall systems in blocks of flats above 18m. The External Wall Fire Review process (EWFR) requires buildings to undergo a fire safety assessment by a suitably qualified and competent professional. At the end of a process, if no action is required, the assessor will sign an EWS1 form to provide formal assurances on fire safety. This is valid for five years. The EWS1 form was originally intended to assess the safety of buildings over 18m tall. However, further advice published in January 2020 has brought many more buildings into scope 'to ensure that the external wall system installed on a residential building meets an acceptable standard of safety, irrespective of height.'[19] This may have unintended consequences, though: some risk-averse lenders may, pending the issue of an EWS1, withhold the granting of a mortgage for flats in blocks, even where the risk is clearly very low. Moreover, it is conceivable that building owners may not permit a survey for fear of unaffordable remediation costs, leaving leaseholders stuck in flats they cannot sell. The process has been criticised by a parliamentary select committee for being unacceptably slow. This inertia in the system can be traced to a shortage of competent and suitably insured assessors to assess and sign off a building to the satisfaction of a mortgage lender.[20] At the current rate of progress, it will take decades to assess all buildings that fall into scope. Meantime, residents have no option other than to endure the potentially serious financial and mental health impacts that further delay will cause. There are an estimated 1.5 million modern flats that are currently unable to be mortgaged, as there is no confirmation that the materials used on the walls are safe. This leaves 3.6 million people unable to sell.[21]

8.20

On 21 November 2020, the Government announced that owners of flats in buildings without cladding will no longer need an EWS1 form to sell or re-mortgage their property. This will allow around 450,000 to sell, move or re-mortgage their flats.[22]

8.21

Where the fire safety standard of external wall cladding on buildings cannot be declared as safe, some leaseholders are facing large bills for additional fire safety measures. These include the installation of alarms systems, sprinklers and temporary waking watches, where immediate evacuation is required in the event of a fire. Insurance costs have increased sharply in some cases. Often leaseholders are left with the uncomfortable choice of finding the money to pay huge premiums or having no building insurance whatsoever.[23]

Property licensing and regulation of larger blocks of flats and student accommodation

8.22

Chapter 6 explored the various property licensing schemes available to regulate the occupation and management of blocks of flats in England and Wales, including parts of the privately managed national stock of purpose-built student accommodation (PBSA). Over the past decade, LHAs have increasingly used licensing schemes to regulate parts of the private rented sector, including flats. These schemes focus on building safety and property management standards. LHA action to reduce the fire risk to all residents and to improve the overall management – particularly fire safety management – must be proportionate and justifiable.

8.23

In recent years, there has been a significant increase in the number of international students registering for full-time courses at UK higher education institutions. This growth has been a major factor in the rapid increase in the number of PBSA blocks. Much of this is new-build, but some of it is converted (for example, from former office blocks, using permitted development rights to meet the demand). According to the recent *Student Accommodation Report*, '*there are now 60,000 beds in the*

UK student accommodation market with over 32,000 new beds entering the market across 40 locations for 2019/20.'[24] Most, if not all, of these new bed spaces are probably in PBSA developments.

In England and Wales, the regulation of student blocks in England varies according to who manages it. LHAs have discretion over what buildings not to include as part of additional HMO licensing schemes, for example, student accommodation where the building management satisfies the terms of an approved code of standards (see Para 6.19). This exemption suggests that the membership of an approved code demonstrates satisfactory fire safety management across the building to permit the exemption of shared flats in multiple occupation from an additional HMO licensing scheme. English LHAs have, though, no powers to except a one- or two-bed studio flat from a selective licensing scheme.[25] In a mixed tenure building containing both shared student flats and studios under the same management, the flats may be exempted from additional licensing. But the studios must be licensed under a selective scheme, where the LHA has introduced such a scheme. The fire safety provisions and management will, however, be identical across all parts of the building, although the risk of a fire is probably higher in the shared flats than in the studio apartments as a consequence of the higher numbers of students. Arguments have been adduced to instigate a review of this apparent anomaly. Disproportionate licensing will divert limited resources away from the worst buildings where action may well be warranted.[26]

8.24

Selective licensing schemes can be targeted to improve the very worst managed properties. These may include some student blocks, although this is unlikely where good management has been independently verified. LHAs are only entitled to impose selective licence conditions to regulate the 'management, use or occupation of the house concerned,' for example, the regular testing of an existing fire alarm system.[27] However, they cannot require the installation or the alteration of a new or existing system, as clarified by the Court of Appeal in *Brown v. Hyndburn*.[28]

8.25

Private providers of student accommodation who manage blocks of shared flats must meet the HMO Management Regulations for each flat. This includes the shared parts of the individual flats, something that is not always understood. The frequency of management inspections may be different from flat to flat, depending on the risks presented by the occupiers. Regular inspections and communication with the students can help reduce the likelihood of accidental ignition, for example, from carelessly discarded cigarettes, candles or poorly located or overloaded electric heaters.

Management of communal parts

8.26

The communal parts of blocks of flats and student accommodation blocks must be regularly inspected and managed in order to:

- reduce the fire risk to occupiers, managers and visitors;
- identify common problems, for example, rubbish stored in corridors, bikes and buggies causing partial blockages, damaged or broken fire doors across escape routes and defective escape route lighting.

Where blocks have been installed with communal fire alarms, fire extinguishers and sprinklers, these must be regularly tested, serviced and maintained by the block managers, who should maintain appropriate records as an audit trail. Fire risk assessors should consider the regularity and effectiveness of common parts management inspections as a measure of confidence in overall building management.

8.27

Block management employees, contractors and residents all have a responsibility to maintain a safe building environment. Effective management communication helps in getting residents to acknowledge their responsibilities and take them seriously. Fire safety action plans should be displayed and kept up to date and relevant, so that residents can readily understand how their actions may have an impact on fire safety in the building and develop the confidence to raise any safety concerns. The engagement of residents is an important part of block management. They should feel assured about the standard of fire safety management, although this may not always be the case. Tower Blocks UK and the University of Kent Law School have produced a residents' fire safety checklist.[29] The purpose of the checklist is to help occupiers identify fire safety failings in the common parts of their individual flats. It also serves as a useful resource for residents to compile evidence for referral to the LHA or fire and rescue service, as part of a request for an inspection. Where resident handbooks and video presentations are provided, they should be updated regularly. This is particularly important in student accommodation.

8.28

Adequate security systems must be provided to prevent unauthorised entry by intruders and to minimise arson, particularly in blocks without a full-time management presence. Where possible, main entry doors to blocks should be self-closing, strong and inspected periodically. Door entry system codes should be changed regularly. Residents should be vigilant and, where possible, prevent unauthorised entry gained by tail-gating. Final exit routes should be well lit and available for use at all times. Under no circumstances should exit doors be locked.

Notes

1 See UK Government Fire Statistics table: *Dwelling Fires Attended by Fire and Rescue Services in England*. The years run from 1 April to 31 March, www.gov.uk/government/statistical-data-sets/fire-statistics-data-tables

2 For further description, see University of the West of England, *Domestic Architecture 1700 to 1960 – Post-War Housing, 1945–1960s – Flats*, https://fet.uwe.ac.uk/conweb/house_ages/flypast/section12.htm

3 See Historic Hansard, *Gas and Electricity: Price Increases*, House of Lords debate, 30 January 1980, vol. 404, cc845–952, https://api.parliament.uk/historic-hansard/lords/1980/jan/30/gas-and-electricity-price-increases

4 There is a difference between the term *common parts* as used in the Housing Act 2004 and the definition in the Regulatory Reform (Fire Safety) Order 2005, which describes areas 'used in common by the occupants of more than one such dwelling.'

5 See statement by West Yorkshire Fire and Rescue Service, *Fines Totalling over Half a Million Pounds Handed Out over 'Potentially Catastrophic' Fire Safety Failings at Leeds Student Flats*, 31 January 2020

6 See Housing Act 2004 sections 1(4)(d) and (5)

7 UK Upper Tribunal decision, *Southwark Council v. Various Lessees of the St Saviours Estate*, [2017] UKUT 10 (LC)

8 See Inside Housing, *Grenfell: The French Connection*, 13 December 2017

9 See Inside Housing, *Grenfell Inquiry: ACM Cladding was 'Primary Cause of Fire Spread' and Tower Did Not Comply with Regulations, Judge Rules*, 30 October 2019

10 See Inside Housing, *Bolton Student Accommodation Involved in Fire Clad with HPL, Planning Documents Say*, 18 November 2019

11 MHCLG, *Advice for Building Owners of Multi-storey, Multi-occupied Residential Buildings*, s.5, January 2020

12 See MHCLG, *Housing Health and Safety Rating System Operating Guidance – Addendum for the Profile for the Hazard of Fire and in Relation to Cladding Systems on High-rise Residential Buildings*

13 See Environmental Health News, *Tales from the Front Line: 'We Expect to See More Blocks Made Safer,'* July/August 2020

14 Inside Housing, *Barking Fire: Risk Assessment Identified 'Significant Risk' from Wooden Cladding Months Before Fire*, 19 June 2019

15 See MHCLG, *Private Sector ACM Cladding Remediation Fund: Full Fund Application Guidance*, July 2019

16 See MHCLG Press Release, *New £1 Billion Building Safety Fund to Remove Dangerous Cladding from High-rise Buildings*, 26 May 2020
17 MHCLG, *Building Safety Programme: Monthly Data Release of a Summary of Latest Figures of Remediation Works to Remove and Replace Unsafe Aluminium Composite Material (ACM) Cladding Systems*
18 SI 2018/1230, *Government Guidance Document Guidance Building (Amendment) Regulations 2018: Frequently Asked Questions*, updated 21 January 2020
19 MHCLG, *Advice for Building Owners of Multi-storey, Multi-occupied Residential Buildings*, https://assets.publishing.service.gov.uk/government/uploads/system/uploads/attachment_data/file/869532/Building_safety_advice_for_building_owners_including_fire_doors_January_2020.pdf
20 See UK Parliament, Housing, Communities and Local Government Committee, *Cladding: Progress of Remediation: The Toll on Residents*, 12 June 2020 https://publications.parliament.uk/pa/cm5801/cmselect/cmcomloc/172/17206.htm#footnote-015-backlink
21 See Inside Housing, *End Our Cladding Scandal Campaign Relaunches with 10-step Plan to Tackle Mounting Crisis*, 27 September 2020
22 Press release, *Government Steps in to Help Homeowners Caught Up in 'EWS1' Process*, www.gov.uk/government/news/government-steps-in-to-help-homeowners-caught-up-in-ews1-process?utm_source=HOC+Library+-+Current+awareness+bulletins&utm_campaign=e4faf02624-Current_Awareness_Social_Policy_I_23-11-2020&utm_medium=email&utm_term=0_f325cdbfdc-e4faf02624-103780058&mc_cid=e4faf02624&mc_eid=cd2f1adce4
23 See The Guardian, *Fire Chiefs Step into Row over Soaring Insurance Costs for High-rise Flats*, 6 October 2020
24 David Feeney (Cushman and Wakefield), *The Student Accommodation Annual Report*, December 2019, www.cushmanwakefield.com/en/united-kingdom/insights/uk-student-accommodation-report
25 Only licences or tenancy agreements issued by specified educational establishments can be excepted from a selective licensing scheme under the Selective Licensing of Houses (Specified Exemptions) (England) Order 2006
26 See MHCLG, *An Independent Review of the Use and Effectiveness of Selective Licensing*, June 2019, updated September 2019
27 Housing Act 2004, s.90
28 See Jonathan Manning, *Selective Licensing and Licence Conditions – The Court of Appeal Gives a Narrow Interpretation to Local Authority Powers*, 4 April, 2018, www.londonpropertylicensing.co.uk
29 See Tower Block UK Fire Safety Checklist, www.towerblocksuk.com/firesafetychecklist

Further reading

House of Common Library, *The External Wall Fire Review Process* (EWS), https://commonslibrary.parliament.uk/the-external-wall-fire-review-process-ews/

9 Future legal developments

Introduction

9.01

Chapter 2 described some of the historical changes made to UK legislation as a consequence of major building fires. It is anticipated that the Grenfell Tower tragedy will also leave a major legacy in the statute books. Towards this end, some heavy machinery of governance has been mobilised:

- a public inquiry into the fire, chaired by retired judge the Right Honourable Sir Martin Moore-Bick;
- an independent review of building regulations and fire safety by Dame Judith Hackitt.

9.02

In the final report of the independent review of building regulations, *Building a Safer Future*, published on 17 May 2018, Dame Judith described the regulatory system covering high-rise and complex buildings as not 'fit for purpose' and concluded that 'a culture change is required.'[1] Her review exposed multiple failings, including ignorance of the building regulations and guidance, excessive focus on quick and cheap in the building of new homes, lack of clarity of roles and responsibilities and inadequate regulation. As a result of these processes, new legislation was required to address the recommendations.

9.03

The Grenfell Tower Inquiry Phase 1 report was published in October 2019 and contained 12 recommendations made primarily to the UK Government, which has indicated its broad in-principle acceptance but has not as yet moved to shape draft legislation.[2] Of the recommendations made, the following are of relevance to building managers, residents and risk assessors:

- information to be made available to fire and rescue services about the materials and methods of construction used in the external walls of high-rise residential buildings. This information will be included as part of the fire risk assessment or other external survey to be conducted by a competent person and made available as required;
- landlords and managers to make and keep under review evacuation plans of HRRBs, and to prepare personal emergency evacuation plans (PEEPs) for all residents whose ability to self-evacuate may be compromised. It is likely that many buildings will continue to operate a stay-put policy, especially where there is no full-time, permanent management presence to coordinate a full evacuation. However, student blocks are more likely to continue to manage simultaneous evacuations;
- the inspection of fire doors and self-closing devices and the regular inspection of fire-fighting lifts to be undertaken.

9.04

At the time of writing, two significant pieces of legislation, the Fire Safety Bill (FSB) and the Building Safety Bill (BSB), are progressing through UK Parliament. These are subject to extensive consultation across England and to parliamentary scrutiny. The bills will tighten fire safety regulation and give residents confidence in the safety of their homes, particularly in high-rise buildings. A new role, Building Safety Regulator (BSB), will be created within the Health and Safety Executive (HSE). Both bills have Opposition support and are anticipated to have a smooth passage through to enactment. When they become law, they will provide for the creation of secondary legislation (regulations) to be issued by the Secretary of State. The new regulations will set out and define exact roles and responsibilities.

9.05

To be effective, it is essential that both sets of legislation are fully aligned, as there will be a responsible person (FSB) and an accountable person (BSB) with similar responsibilities for fire safety. In many buildings they are likely to be the same person. EHOs will continue to assess buildings and to work closely with fire and rescue authorities. But they will also engage with the new regulator within the HSE:

> *The FSO and the Housing Act 2004 (where appropriate) will continue to apply alongside the Building Safety Bill and the Government intends to address the interaction between the different regimes within buildings in scope of the new building safety regime by ensuring that regulators provide stakeholders with comprehensive operational guidance.*[3]

Fire Safety Bill (FSB)

9.06

On receiving Royal Assent, the FSB will become the Fire Safety Act 2020 ('the Act'), limited to England and Wales only. It will amend the Regulatory Reform (Fire Safety) Order 2005 (FSO) to clarify that the FSO extends to the structure, external walls and residential flat front doors in any building containing more than two domestic premises, irrespective of height.[4] External walls also include 'doors or windows in those walls' and 'anything attached to the exterior of those walls (including balconies).'[5] It will complement:

- the existing HHSRS powers available to LHAs to take enforcement action against building owners and managers;
- the Building (Amendment) Regulations 2018.[6]

The FSB removes the ambiguity of whether individual flat front doors and the external structure are included in the scope of FSO. Enforcement of the Act will continue to fall to the fire and rescue authorities.

9.07

There has been general industry-wide acknowledgement that the Act will increase the burdens on the responsible person of multi-occupied residential buildings that fall into scope. Landlords and managers will have to reach clear agreement on who the responsible person is. It is likely that increased costs will be passed on, at least in part, to the individual leaseholders through increased service charges. Depending on their terms, individual leases may increase rents, especially in the private rented sector.

Competence of fire risk assessors

9.08

It is widely anticipated that larger numbers of competent fire risk assessors will be needed to undertake the expected increase in assessments – especially in HRRBs:

> *The Government proposes to amend the FSO to require that any person engaged by the RP [responsible person] to undertake all or any part of the fire risk assessment must be competent.*[7]

The costs of a risk assessment may also increase if insurers and mortgage companies request a document from a 'competent person,' who in turn may need to have adequate liability insurance. There are very many more assessors who are competent to undertake an assessment of a two-storey block of flats than a 30-storey tower block. Further Government guidance will be required as the legislation is formulated to clarify what is meant by the term *competent* and how to achieve sufficient competence to undertake an assessment of a taller building.

9.09

The Competence Steering Group (CSG) has recommended that fire risk assessors for HRRBs should be suitably third-party accredited and registered.[8] It has also advised that only accredited assessors should be engaged to assess larger HRRBs.

Inspection and verification of fire doors

9.10

Sir Martin Moore-Bick recommended that fire doors in all buildings, irrespective of height, should be inspected every three months by the owner or manager (responsible person) to ensure the self-closing devices are effective and all fire doors provide satisfactory fire resistance. The Government is seeking to reduce the time between inspections to intervals of three to 12 months, depending on the height of buildings and the location of the doors.[9] Under the recommendations, only fire doors in communal parts of high-rise blocks would require inspection every three months. In mixed tenure blocks containing owner-occupied and private rented flats, the responsible person would have to ensure ongoing resident engagement to obtain access for inspections and verify the front doors to flats.

9.11

The Act will confirm that the front doors to flats will fall within the definition of common parts and become the landlord's responsibility. Residents will be expected to co-operate with the responsible person. It is possible, though, that a conflict may arise over the ownership of the door, depending on the terms of the lease, especially where a leaseholder has replaced the original front door. The Upper Tribunal has examined the terms of a lease where a leaseholder replaced two original flat front doors with suitable fire-resistant doors. The landlords sought to replace these doors, citing a breach of the lease, and to re-charge the leaseholders for the cost. The Tribunal found in favour of the leaseholder, and the landlord was prevented from changing the doors.[10] Landlords should check the terms of the lease to clarify the legal status of leaseholder flat door ownership to avoid conflict and possible litigation.

Building Safety Bill (BSB)

9.12

The Government presented the draft Building Safety Bill to Parliament on 20 July 2020 as a way to implement the recommendations of the Hackitt Review. The draft Bill will be extensively scrutinised before it is introduced to Parliament in full. The explanatory notes to the Bill confirm that the draft legislation will include the creation of the role of Building Safety Regulator, whose functions will be to:

- implement a more stringent regulatory regime for higher-risk buildings. The draft Bill does not define 'higher-risk buildings.' However, the proposal would bring into scope buildings over 18m and buildings that have more than six storeys;
- oversee the safety and performance of all buildings;
- assist and encourage competence in the built environment industry and registered building inspectors.[11]

9.13

Residents with concerns will be able to lodge complaints with an 'accountable person,' who, as duty-holder during the building's occupation, will be legally responsible for:

- registering all buildings in scope with the BSR;
- knowing and understanding the fire and structural risks in the building;

- taking appropriate steps to mitigate and manage these risks so the building can be safely occupied.

As part of their duties, the accountable person will have control of the common parts of a building and will be required to assess the building safety risks and produce and keep up to date a Residents' Engagement Strategy.

The Building Safety Manager's role will be to support the accountable person in the day-to-day management of fire and structural safety in the building. A key driver here is effective communication to keep residents and leaseholders safe in their homes.

Extension of sprinklers in residential buildings

9.14

In Melbourne, the full evacuation of Lacrosse Tower was made easier by sprinklers working in combination with an automatic alarm system. There was no such requirement for sprinklers to be installed in Grenfell Tower, as it was built in 1974. Sprinklers were not retrofitted as part of a refurbishment completed in 2016, even though the retrofitting of sprinklers in HRRBs was recommended by the coroner following the inquest into the Lakanal House fire in 2013.[12] Sprinklers fitted in tower blocks can reduce loss of life: in 2017, BBC Scotland reported that *'fifteen people have died and more than 480 have been hurt in high-rise fires in Scotland since 2009 . . . just one of those casualties was injured in a flat fitted with a sprinkler system.'*[13]

9.15

The requirement for sprinklers in residential buildings across the UK presents a mixed picture. They were introduced into the Building Regulations in Scotland in 2005[14] for high-rise blocks of over 18m and in England in 2007 for new-build blocks taller than 30m.[15] As the regulations are not retrospective, many older tall buildings still do not have sprinklers installed for a variety of reasons. By 2019 an investigation revealed that, in England, many tall buildings did not have sprinklers: 'among 2,107 council-owned tower blocks taller than 10 storeys, only 112 had sprinklers [fitted].'[16]

9.16

In England, the threshold for installation of sprinklers in new-build and materially altered blocks of flats and in mixed-use buildings containing flats will be reduced from 30m to 11m from November 2020.[17] The

Scottish Government introduced regulations that came into force on 1 March 2021.[18] These will require all new apartments, new shared multi-occupancy residential buildings and all new social housing to be protected with sprinklers. Beyond this date, only new-build single family houses in Scotland will be excepted from the requirement to have sprinklers.

International fire safety standards

9.17

There are many different approaches to fire safety standards around the world, in both construction, design and building management. The International Fire Safety Standards Coalition (IFSSC), a group of over 80 fire safety organisations, was launched on 9 July 2018 in the wake of the Grenfell Tower fire. The Coalition's primary aim is to encourage a consistent approach to reducing fire risk in buildings and loss of life, and to share knowledge globally.

9.18

On 5 October 2020, the IFSSC published *International Fire Safety Standards: Common Principles* (IFSS-CP), which lays a basis for achieving a more consistent approach to fire safety internationally.[19] The IFSSC will prepare a universal set of fire safety classifications and definitions at project, state, national, regional and international levels. A key objective for the Coalition is for all higher-risk buildings to which occupiers and the public have access to meet IFSSC standards and to publicly display a certificate of compliance.[20]

Notes

1 Dame Judith Hackitt, *Building a Safer Future: Independent Review of Building Regulations and Fire Safety: Final Report*, May 2018, https://assets.publishing.service.gov.uk/government/uploads/system/uploads/attachment_data/file/707785/Building_a_Safer_Future_-_web.pdf

Dame Judith Hackitt's personal view is that 'the above issues have helped to create a cultural issue across the sector, which can be described as a "race to the bottom" caused either through ignorance, indifference, or because the system does not facilitate good practice. There is insufficient focus on delivering the best quality building possible, in order to ensure that residents are safe, and feel safe.'

2 Grenfell Tower Inquiry: Phase 1 Report Overview, https://assets.grenfelltowerinquiry.org.uk/GTI%20-%20Phase%201%20report%20Executive%20Summary.pdf

3 See Home Office, Fire Safety, Government Consultation, *Alignment with the Building Safety Bill*, July 2020, https://assets.publishing.service.gov.uk/government/

uploads/system/uploads/attachment_data/file/919566/20200717_FINAL_Fire_Safety_Consultation_Document.pdf

4 SI 2005/1541

5 House of Commons Library Briefing Paper Number 8782, 3 September 2020

6 SI 2018/1230

7 See Home Office, Fire Safety, Government Consultation 1.3, *Quality of Fire Risk Assessments*, July 2020, https://assets.publishing.service.gov.uk/government/uploads/system/uploads/attachment_data/file/919566/20200717_FINAL_Fire_Safety_Consultation_Document.pdf

8 See Competence Steering Group for Building a Safer Future, *Setting the Bar – A New Competence Regime for Building a Safer Future*, October 2020, http://cic.org.uk/admin/resources/setting-the-bar-exec-summary-final.pdf

9 See Home Office, Fire Safety, Government Consultation 2.9, *Fire Doors*, July 2020, https://assets.publishing.service.gov.uk/government/uploads/system/uploads/attachment_data/file/919566/20200717_FINAL_Fire_Safety_Consultation_Document.pdf

10 See *Fivaz v. Marlborough Knightsbridge Management Ltd* [2020], UKUT 138 (LC)-, www.bailii.org/uk/cases/UKUT/LC/2020/138.html

11 See Building Safety Bill Explanatory Notes, 20 July 2020, https://assets.publishing.service.gov.uk/government/uploads/system/uploads/attachment_data/file/901869/Draft_Building_Safety_Bill_PART_2.pdf

12 See letter to Department of Communities and Local Government, 28 March 2013, www.lambeth.gov.uk/sites/default/files/ec-letter-to-DCLG-pursuant-to-rule43-28March2013.pdf

13 BBC Scotland, *Effect of Sprinklers on Fire Safety in Scotland's Tower Blocks*, 27 September 2017

14 Scottish Government, *Building Standards Technical Handbook 2017: Non-domestic Buildings*

15 Building Regulations Approved Document B:2, *Buildings other than Dwelling Houses*

16 See Inside Housing, *No Sprinklers in 95% of Tower Blocks, 10 Years on from Lakanal*, 2 July 2019

17 The Building Regulations 2010, *Amendments to the Approved Documents*, https://assets.publishing.service.gov.uk/government/uploads/system/uploads/attachment_data/file/887210/AD_B_2019_edition__May2020_amendments.pdf

18 Scottish Statutory Instruments 2020, No. 275, *The Building (Scotland) Amendment Regulations 2020*

19 See *International Fire Safety Standards: Common Principles*, https://ifss-coalition.org/wp-content/uploads/2020/10/IFSS-CP-1st-edition.pdf

20 See RICS, *New Worldwide Fire Safety Standard Launched*, 5 October 2020, www.rics.org/uk/news-insight/latest-news/news-opinion/international-fire-safety-standards/

Index

Note: Page numbers in **bold** indicate a table. Page numbers in *italics* indicate a figure.